BIOHACK YOUR BRAIN

BIOHACK YOUR BRAIN

How to Boost Cognitive Health, Performance & Power

DR. KRISTEN WILLEUMIER

WITH SARAH TOLAND

wm

WILLIAM MORROW

An Imprint of HarperCollins*Publishers*

This book is written as a source of information only. The information contained in this book should by no means be considered a substitute for the advice, decisions, or judgment of the reader's physician or other professional adviser.

All efforts have been made to ensure the accuracy of the information contained in this book as of the date published. The author and the publisher expressly disclaim responsibility for any adverse effects arising from the use or application of the information contained herein.

HarperCollins books may be purchased for educational, business, or sales promotional use. For information, please e-mail the Special Markets Department at SPsales@harpercollins.com.

A hardcover edition of this book was published in 2020 by William Morrow, an imprint of HarperCollins Publishers.

FIRST WILLIAM MORROW PAPERBACK EDITION PUBLISHED 2021.

Library of Congress Cataloging-in-Publication Data has been applied for.

ISBN 978-0-06-299433-2

23 24 25 26 27 LBC 7 6 5 4 3

To the awakening of greater health and well-being for all of humanity.

This book was written in loving memory of my parents, who provided an environment that encouraged scientific exploration, creativity, independent thinking, and the pursuit of higher knowledge. My father, a proud marine and firefighter, loved serving his country and community and lived the final decade of his life with a neurological condition. May the memory of his strength, courage, optimism, faith, and perseverance live on throughout the pages of this book and touch the hearts and minds of all who read it.

CONTENTS

FOREWORD

When I met Kristen, I was looking for a neuroscientist who understood cognitive analytics to work on a research trial identifying dementia in asymptomatic patients. Kristen was ideally qualified, having both a broad base of knowledge in cognitive science and the unique experience of running research for a large brain-imaging clinic for years.

Our research team for the trial included many eminent doctors, neuropsychiatrists, and similar sorts. But from the onset, there was something special about Kristen—she was committed, driven, and determined to find solutions, which was all apparent from the moment I met her. Nonetheless, I was still surprised when she brought testing tools into our trial that not even the neuropsychiatrists who regularly relied on them had thought of. These were tools I hadn't thought of, either.

I've spent more than forty years studying and treating the brain. I've conducted numerous clinical trials and have been fortunate enough to be recognized for my work, including on the cover of *Time* magazine's "Heroes of Medicine" issue. You can imagine then how much Kristen impressed me when she showed a cognitive-health vet like me how things might be done differently.

Kristen is absolutely a committed and compassionate spokesperson for the brain. Not many neuroscientists and neurosurgeons can communicate clearly about what we can do to improve cognitive function, but she is concise and compassionate, and she speaks about the brain in a language everyone can understand.

When it comes to the brain, clear communication is increasingly invaluable. Over the past decade, we've seen more and more information disseminated about what we can allegedly do to optimize our cognitive health. Every time you open your news feed, there's another article—what's good, what's bad, what you should take and shouldn't take.

Much of this information, however, isn't based on good scientific data. If you search Google, for example, you'll find hundreds of supplements claiming to improve cognitive function, even though few have quality research supporting these claims. I've often wished I had a handbook I could give patients that outlined the interventions we actually know can improve brain health.

I now have one. *Biohack Your Brain* leverages what research has shown to optimize cognitive health. This is relatively new science and continually evolving, which is why it's critical to get advice from a trusted source. We've only recently learned, for example, that diet, exercise, mindfulness, sleep, and stress regulation play a big role in cognitive function—and in different ways than they do in heart health. And we now know certain lifestyle interventions can slow the risk of developing dementia by as much as a decade.

If you only plan to live until you're forty, this book may not help you. But if you want to live as long and as well as you can, *Biohack Your Brain* can have a significant impact on both your cognitive health and overall quality of life.

In other words, you can biohack your brain if you have the

right resources and information to do so. Personally, I can't think of anyone more qualified to show us the best ways to biohack the brain than Kristen. She's even taught me a thing or two.

—Keith L. Black, M.D., chair and professor, Department of Neurosurgery, Cedars-Sinai Medical Center

BIOHACK YOUR BRAIN

PROLOGUE

The Case for Your Cognitive Health

The gospel of health and wellness is everywhere these days. You hear about the newest trends and solutions espoused by medical businesses, books, websites, food and fitness companies, hospitals, and health conglomerates. They all tell you to join a gym, try a diet, take these supplements, lose weight, lower cholesterol, drop blood pressure, get heart-healthy, prevent cancer . . . the noise can be deafening.

Amid all this, we don't hear nearly enough about the brain—the only organ in the body we can't live without, the one that orchestrates our entire lives.

I've been fascinated by the brain ever since I was a psychology major at Boston College. After earning my bachelor's degree, I was inspired to continue my education, getting a master's degree in physiological science and a doctoral degree in neurobiology from the University of California, Los Angeles.

During my graduate and postgraduate work, I spent years in research labs studying neuroendocrinology, neurophysiology, and neurogenetics. During this time, I was awarded a fellowship from the National Institutes of Health, which opened up opportunities to present my research at conferences around the world.

Following my postdoctoral training, I went into the field of brain imaging, taking the role of research director for the Amen Clinics, a nationally recognized mental health care center for the study of the brain. My time there changed my life—and the lives of many others. It was at the clinic that I helped lead several clinical trials with NFL players that revealed just how much cognitive damage the sport can cause—a groundbreaking conclusion at the time when my colleagues and I published our results. More important, we discovered noninvasive ways to help treat and even reverse some of the damage we saw by using diet, supplements, exercise, and cognitive training.

Later, I was lucky enough to be able to use the knowledge gleaned from those trials to help my father, who passed away in 2017 after years of living with symptoms associated with Parkinson's disease. While it was incredibly painful to watch his condition progress, I was able to hold on to hope knowing that I did have some tools to help him retain his quality of life. As his condition worsened, I encouraged him to adopt some of the techniques we used with the NFL players to help heal their brains. It was incredible to see how his balance and grip improved—such that he was able to hold on to some independence until the very end. To this day, I'm so heartened that I was able to give him some tools to make his last years more pleasant and full of hope.

This story epitomizes the number one lesson I've learned in my twenty-plus years of study: everyone has the potential to change his or her brain. No matter how old you are or the choices you've made in the past, it is always possible to make improvements.

Today, after all, millions of Americans contend with cognitive issues. Many blame symptoms like memory loss, brain fog, poor concentration, anxiety, and depression on physical problems when the real root is brain based. It can be easy to dismiss cognitive problems as the by-product of stress—which, admittedly, is everywhere. Still, that doesn't mean you have to let it derail your cognitive performance. There are several science-backed ways to release stress's grip on the brain in order to restore your cognitive power and potential.

If you're concerned with dementia, you have every right to be. The umbrella term for cognitive impairment currently impacts 10 percent of all Americans over the age of sixty-five—a statistic expected to grow as the country ages. Cellular changes that lead to dementia can occur decades earlier, even as early as your thirties and forties, which is when the brain begins to age. This makes it the perfect time to care for your brain now, no matter how old you are, leveraging the time when you can make changes in your brain health habits in order to circumvent a dementia diagnosis later in life.

If you've experienced a mild brain injury like a concussion or are worried what the trauma could do to your children or grandchildren, especially if they play sports, the best way to confront fear is with knowledge. A brain injury isn't a reason to give up, but an opportunity to learn what you can do with your diet, exercise, supplements, and other lifestyle choices to restore your cognitive health.

The coronavirus crisis has also made taking care of your brain more imperative than ever before. The global pandemic has raised the stress, anxiety, and fear of millions of Americans, sabotaging their mood and interfering with their cognitive function and health. In this book, you'll learn how to counter that kind of stress, fear, and negative emotion in order to boost your brain and strengthen your mental resolve against a similar sort of tragedy in the future. What's more, everything you'll discover about how to improve your cognitive performance—what to eat, how to exercise, which supplements to take—will also improve your immune function, bolstering your body in the advent of another outbreak.

I'm writing this book to show you that no matter what the current state of your brain health is, you have the power to biohack your brain and improve its function. If you've ignored your cognitive health for years, you have the potential to turn this promise into a reality. I know because I've seen it happen, over and over again, even in those who have suffered debilitating degrees of cognitive damage. After all, if football players who've spent years taking hard hits to the head can change their brain in a matter of months, you can, too.

This is all to say that everyone has his or her own journey, and my purpose is to help you discover what you can do along that journey to harness the true power of your brain. Instead of thinking of the brain as an abstract structure inside your head, I'll show you how this incredible organ orchestrates your physical movements, directs your conscious mind, and powers the inner workings of your intelligence and personality. In short, your brain is what makes you *you*—unique, beautiful, and blessed to be alive.

1

YES, YOU CAN CHANGE YOUR BRAIN

I live in Los Angeles. If you've ever been, you know the weather is warm, the beaches are beautiful, and the cars that drive down our palm-lined boulevards are incredible.

I'm not necessarily a car person per se, but when you live in the City of Sunshine, you can't help but be charmed by L.A.'s car culture. Spend an afternoon on Santa Monica Boulevard and you'll see everything from classic Cadillacs and old Aston Martins to brand-new Teslas, Porsches, and Ferraris. We also have plenty of beaters in L.A., including cars you can barely identify by model or even manufacturer because the hood is so crumpled, the back is banged up, and the exterior has been painted several times.

The reason I'm carrying on about L.A.'s car culture at the beginning of a brain book is because I can't think of a better analogy to explain the brain and why it matters how we take care of it. Similar to how a car is a complicated piece of machinery with hundreds of different working parts, so is the organ inside our heads that controls nearly every cell, thought, and behavior. Each and every part of our brain matters, as it does in a car: Let an inner valve rust, blow a fan deep inside the engine, or ding up a pump you've never even heard of, and your car may not run as well. Well, your brain is exactly the same.

I think about this car-brain analogy every time I step inside my front door because Mark, my fiancé, restores classic cars as a hobby. Our garage is full of old beauties, and he's won multiple awards that line our hallways and home office. I've learned from him, along with living in L.A.'s car kingdom, that people who are passionate about taking care of their automobiles are able to do amazing things with them, maintaining models from the 1950s, '60s, and '70s, so they look and run as well as any contemporary car. Conversely, people who don't take care of their cars often end up having to rely on something that doesn't handle well, is too slow, can't last as long, or may even be unsafe.

Here's where the car-brain analogy ends, though. Your brain, unlike your car, is a living, oxygen-consuming supercomputer with extraordinary processing capabilities. More than that, it's an essential part of who you are. So the consequences of not caring for your brain can be far worse than an expensive repair bill or getting stranded on an empty highway. If you don't perform regular maintenance on your brain by adopting brain health habits, you can jeopardize your ability to generate new ideas, maintain your focus, learn new information, and remember all the little things in life that make our time on this planet so precious. Without continual brain care—the automobile equivalent of changing the oil, replacing the fluids, checking the tire pressure, and swapping out old air filters and dead batteries—your brain won't last as long, physically or mentally, and won't run as efficiently. And while you can always get a different car if you end up wrecking the one you have, or trade up if you lease, you can *never* get a new brain.

Not only are you stuck with one brain for life, it's also the most essential organ in your body when it comes to physical, mental, and emotional operations. Our brain controls everything we do, both our intentional actions—what we say, how we say it, how we move, and whether we want an ice cream sundae or kale salad—and the automatic ones, like our heart rate, blood pressure, breathing, sleep cycles, hunger, and thirst.

It also interprets and translates sensory information from the rest of your body, controlling what you see with your eyes, what you hear with your ears, what you smell with your nose, what you touch with your skin, and what you taste with your tongue.

The brain also communicates directly with other areas of your body, sending and receiving millions of messages through your spinal cord, which, along with the brain, make up the central nervous system. The central nervous system collates the body's physical and sensory information and coordinates physical, mental, and emotional activities across your entire body.

Just as the brain is the most important organ in our body, it's also the most complex. Our brain contains approximately 100 billion brain cells, known as neurons, and billions more glial cells, which support the neurons. A single neuron can form thousands of connections with other neurons, using gaps known as synapses to send messages between cells. This incredible labyrinth of elaborate cells, conduits, and signals results in more than 100 trillion connections in our brain—one reason why the human brain has been called "the most complicated object in the known universe."[1]

The good news? I'm here to help you crack the code behind biohacking your brain.

Changing the Brain Isn't Brain Science: What I Learned at a Nationally Recognized Brain-Scan Center

While the brain is immensely complex, the ways we can change the brain aren't nearly as complicated. In fact, changing your brain is really easy! After completing my doctoral and postdoctoral training, I started working as the director of research for Amen Clinics, where physicians treat a panoply of physical, mental, and emotional issues using information from patients' clinical histories and brain scans. I was surprised to see firsthand just how effective small changes to our daily routines, practiced consistently, can be to optimize brain health. These lifestyle changes can be as simple as choosing one food over another, engaging in a specific type of exercise, adopting a different mental approach to a common daily situation, and following certain protocols that are simple enough for a fifth grader to understand.

The clinics minister to a full range of cognitive conditions, including the kind you'd expect to see: dementia, Alzheimer's, memory issues, and other neurodegenerative problems. They also treat mental health issues like anxiety, depression, attention-deficit hyperactivity disorder (ADHD), self-harm, and suicide, anger management, schizophrenia, obsessive-compulsive disorder, bipolar disorder, and borderline personality disorder. Some patients have suffered concussions or other traumatic brain injuries, while others have illnesses that affect the entire nervous system, like Lyme disease or toxic mold exposure. This range of problems is treatable, though, using information from patients' brain scans to help tailor lifestyle choices like diet, exercise, and supplements that will help influence their cognitive function and health.

One of the top issues we helped patients manage at the

clinics was weight loss, since excess body fat has severe consequences on brain health. I coached hundreds of people on how to lose weight and keep it off using simple lifestyle protocols based on brain data.

My experience as the research director leading clinical neuroimaging trials gave me the equivalent of a de facto Ph.D. in the premise of this book: what the everyday person can do to biohack his or her brain. I saw thousands of brain scans both before and after patients implemented treatment protocols, and I was awed by remarkable and even inspirational differences they were able to make in a few months' time with simple lifestyle modifications.

A profound instance of this occurred when I helped lead a clinical research trial on current and retired football players in 2009. At the time, there had never been a large-scale study using brain imaging in living players to understand, in a comprehensive way, what was really happening under the helmet. For the study, we recruited one hundred active and retired NFL players from twenty-seven teams across all positions. To participate in the study, all of these guys had to have spent at least three years active on an NFL roster—in other words, these players weren't the ones keeping the bench warm. Many, both the offensive and defensive players, had experienced multiple hard hits and concussions, in addition to hundreds, if not thousands of milder subconcussive impacts.

While we were expecting to see some brain trauma, we were shocked to uncover the degree of damage in the players, who were some of the best athletes in the world. These players were highly tuned and conditioned—or they had been. They had spent their whole lives training, sleeping, lifting, eating, and breathing with a single goal in mind: to play and win the game of football. Technically, their brains should have been relatively healthy, not some of the unhealthiest the clinic had ever encountered.

We first put the players through a comprehensive set of neuropsychological and neurocognitive tests and conducted functional and electrical brain imaging. This allowed us to look deep inside their brains to see which areas were working well and which were not functioning as optimally as we would like them to. What we saw was eye-opening. Most players weren't getting the blood flow to their brains they needed, especially in areas responsible for memory and basic cognitive functioning.

While startling, the players' brain scans weren't discouraging. We believed we could help them recover their cognitive function and give them back the wonderful things their brains could once do quickly and effectively, both on and off the field. But doing so would mean changing their daily routine, which meant earning their trust.

For the next six months, we talked with the players, taught them about their brain function, and asked them to make specific lifestyle and dietary changes based on their personal cognitive data. Each player's individualized protocol dictated when and how much he slept and which nutritional supplements he should take and avoid. I coached them through it, meeting with them often in groups or one-on-one and cheerleading them to stick to the protocol. In the end, this earned me the nickname "Coach K."

After six months, we rescanned the players' brains and administered the same tests we did when they first arrived. What we saw then was even more impressive than their initial scans. In just 180 days, these men, who once had some of the unhealthiest cerebral perfusion, or blood flow, we had seen, had turned around their brain function. In their six-month scans, we could see clearly that their brains were better perfused with blood and that they had restored function to certain cognitive areas previously damaged by poor health and repetitive hits.

If professional football players can change their brains, anyone can—and you'll have an easier time changing your brain unless you've also taken multiple hits from 250-pound men wearing twenty pounds of gear and polycarbonate helmets.

The Three Most Significant Ways You Can Change Your Brain

1. Yes, You Can Grow New Brain Cells at Every Age

First, a little truth: we lose thousands of brain cells every day as part of the natural aging process. Some of us lose more brain cells than others due to too much stress and exposure to heavy metals, pesticides, and other toxic chemicals in our environment, water, and food. Of course, having a drug and alcohol problem or suffering a mild brain injury, stroke, or cognitive disease like Parkinson's or Alzheimer's can also cause brain cell loss.

Now, a little good news: Our brain contains approximately 100 billion neurons, or brain cells, which are some of the longest-living cells in the body. The vast majority of the neurons we're born with and develop as a child remain with us for our entire lives—it's why maintaining your neuronal health is critical to your long-term cognitive function.

Finally, some fantastic news: Scientists used to think we couldn't grow new neurons as adults, but as it turns out, they were wrong. You can produce new brain cells as you age, even if you're in your sixties, seventies, or eighties.

The process of growing new brain cells is called neurogenesis. It takes place in the area of our brain known as the hippocampus, a seahorse-shaped structure located deep inside the brain's inner regions that plays a major role in memory and learning. You'll get to know the hippocampus much better in chapter two: "Brain Basics."

Neurogenesis isn't only for athletes or young people eager to unlock optimal brain function. Recent studies show that people in their seventies, eighties, and even nineties can stimulate new neuron growth by changing their exercise, diet, stress, sleep, and supplement habits. Research even shows older people, including those with Alzheimer's, can grow as many new neurons as young people.

When you create healthy new cells, you improve your neural capacity to activate, connect, and respond to all the information your brain processes and receives. The more healthy cells you have, the more quickly and effectively you can make smart decisions, hone your focus, preserve your memory, and retain facets of what we call our executive function—a broad umbrella term for the higher-level cognitive skills that control our behavior. Since neuronal death is the hallmark of brain aging, the more you're able to slow or counter that process with new cell growth, the younger your brain will be.

More specifically, studies show neurogenesis increases the volume and function of the brain's hippocampus, which can help preserve and even boost memory and learning. Growing new brain cells also helps you to better deal with stress and can help mitigate mood disorders like depression, anxiety, and even posttraumatic stress disorder. And while the research is still in the early stages, studies look promising that brain-cell growth in the hippocampus can also play a role in slowing or even reversing the progression of cognitive diseases like Alzheimer's.

Both neurogenesis and neuroplasticity, or changes in neuronal connections from new learning, demonstrate the brain's ability to change over a person's lifetime. Growing new neurons is one way we can remodel our brain, giving us the lifelong ability to improve our cognitive function.

Throughout this book, you'll learn specific, science-backed ways to trigger new neuron growth. These modifications include certain forms of exercises, foods and nutritional supplements, and adaptations to how you handle stress. Some modifications are specific—for example, not all types of exercise have been shown to instigate new neuron growth. And similar to a car, the kind of consistency you give these habits makes the difference between a just-functioning brain and a shiny, new restoration job.

2. It's All About Blood Flow, Baby

This may not sound like fancy neuroscience, but boosting blood flow to the brain is exactly what evidence-based research shows we need for optimal cognitive health and performance.

If you're thinking that sounds simple, you're right. But simple doesn't mean universal: most people don't have optimal cerebral circulation.

The reason suboptimal cerebral circulation is so widespread requires understanding two components of brain health. First, our brain needs a rich and steady flow of blood to function properly. Second, many modern-day lifestyle habits negatively impact cerebral circulation without causing symptoms or problems until it's often too late.

While our brain takes up only 2 percent of our total body weight, it requires 15 to 20 percent of the body's total blood supply. Your body will even stop directing circulation to other organs in order to maintain a flow of oxygen and nutrient-rich blood to your cerebral headquarters.

The brain also uses three times as much oxygen as your muscles do. Blood is the only way to get oxygen to your brain cells so that they can function, fire, and signal efficiently. Without proper blood flow, brain cells begin to die.

Blood is also the brain's only source of glucose, or sugar, which brain cells suck up for fuel. Unlike your muscles, the brain can't store glucose, so if you're not getting enough blood to the brain, you're starving your cerebral tissue. And your brain is a hungry organ: it consumes 40 to 60 percent of the body's total blood glucose. Additionally, blood brings other vital nutrients to the brain, including vitamins, minerals, fats, amino acids, and electrolytes.

If you reduce your brain's nutrient and oxygen supply by even a fraction, you also reduce your brain's ability to activate areas that help dictate mood and cognitive function, including the ability to concentrate, remember details, come up with new ideas, make good decisions, and multitask.

There's another crucial role cerebral circulation plays: rinsing away tissue of metabolic waste that can build up over time. This includes the amyloid-beta protein, a protein that can become toxic when accumulated in the brain and has been associated with the development of Alzheimer's disease.

If you're suffering from brain fog, concentration issues, or memory problems, you might blame a number of other issues—poor sleep, stress, bad diet, or maybe an underactive thyroid—before you'd ever think about your cerebral circulation. It's simply not a common aspect that most patients or health practitioners consider.

Why do so many people have suboptimal blood flow? We can blame many of our modern lifestyle habits, including how we eat, drink, sleep, exercise, and deal with everyday stress. While that's an extensive list, modifying just a few habits can go a long way toward optimizing brain health.

3. Calming Your Sympathetic Nervous System Can Change Your Brain Overnight

According to the Mayo Clinic, stress is "a normal psychological and physical reaction to the demands of life."[2] In other words, stress is natural and even has some health advantages. The body's fight-or-flight response—a chain of reactions that occurs whenever we face a suddenly stressful or even life-threatening situation—triggers the production of hormones, chemicals, and brain activity that we need, for example, to run faster if being chased by a predator, fight harder when cornered by an assailant, or lift a two-ton car off a loved one pinned beneath.

In addition to aiding in life-or-death situations, stress has other healthy functions. A modicum of acute stress can help motivate us to take action, hone our focus when we need it to accomplish a task, and help give us a sense of fulfillment or accomplishment after the stressful event ends.

But the key words in this last sentence are *after the stressful event ends.* Otherwise, too much stress, sustained over a period of time, is detrimental to the brain. Prolonged stress slows cerebral circulation by causing plaque buildup that narrows arteries and can constrict or even permanently damage the brain's blood vessels. When we're stressed, our muscles tense, particularly in our neck, further reducing blood flow to the brain.

Chronic stress is also terrible news for your neurons. If your stress levels are elevated for too long, your brain can't engage in neurogenesis and—even worse—can start killing cells. Chronic stress also ages cerebral tissue and can act on your neuronal life span in ways that are similar to a concussion or the beginnings of a neurodegenerative disorder.

The brain cells left alive when you're stressed aren't particularly healthy, either. Chronic stress causes neurons to overactivate. Over time, this can establish new neural pathways that can change how your brain functions.

We can't talk about stress without detailing the hormone largely responsible for many of its detrimental effects: cortisol. Cortisol is produced when we suffer any type of stress, both good and bad. A little cortisol isn't necessarily harmful and can even have some benefits. But too much can be deleterious, causing everything from weight gain and sleep disruptions to causing the hippocampus to shrink, interfering with our ability to concentrate and recall facts and circumstances. At the same time, the hormone can increase the size and activity level of your amygdala, an almond-shaped group of neurons deep inside the brain that helps attach emotional significance to memories. A larger and more active amygdala can make us more sensitive to fear and anxiety.

Chronic stress has other harmful effects for brain health. Stress can create more white matter, the fatty tissue that makes up half our cerebral tissue where many neural connections occur.[3] Too much white matter means less room for gray matter, which is where the brain processes all the body's physical, emotional, behavioral, and sensory information. This imbalance can create emotional and cognitive problems that don't necessarily subside when our stress does.

Most people associate stress with the emotional hardship that comes with a traumatic event like selling a home or managing an illness or injury, or something related to daily stressors, like dealing with work pressure, handling bills, and taking care of children or other family members.

But stress comes in other forms as well. Physical stress can be triggered by diseases like arthritis, diabetes, and dementia and can also be produced by high blood pressure, poor diet, too little sleep, and chronic dehydration. You can also incur chronic stress from working out too much or, conversely, not moving your body enough.

In addition to mental, emotional, and physical stress, you can also be exposed to environmental stress—a growing problem in our modern world, where chemicals are used to manufacture just about everything we eat, drink, wear, put on our skin, and use inside our homes and offices. Air pollution, present in every molecule we breathe, can also increase stress and be especially harmful to the brain, raising the risk of cognitive decline and disease.[4]

I know, you're probably getting stressed considering all the different types of stress! But before we delve into solutions, there is one more important stressor to be aware of: electromagnetic fields (or EMFs). These are "invisible areas of energy, often referred to as radiation," as defined by the National Institute of Environmental Health Sciences,[5] emitted by cell phones, computers, WiFi networks, microwave ovens, hair dryers, televisions, power lines, and other electrical equipment and wireless transmitting devices.

While many cell phone and technology manufacturers claim low-level radiation from their products is harmless, studies show otherwise. Research on cell phones shows EMFs can modify brain excitability, or how likely it is for neurons to fire. Too much brain activity can cause neurons to get overexcited, impairing brain health and function. EMFs can also limit blood flow to the brain, cause memory loss, and even damage neuronal DNA.[6] Evidence also suggests EMFs can interfere with the body's sleep cycle and energy levels, and contribute to the development of other medical problems like weight gain, headaches, dizziness, and even cancer.[7]

Kristy's Story

WHAT STRESS CAN DO TO YOUR BRAIN WITHOUT YOUR REALIZATION

Kristy* came to see me to have her brain health evaluated. Her mother and brother had passed away in their late fifties from a rare form of brain cancer known as

* Some client names have been changed to protect patient privacy.

glioblastoma, and at age fifty-four, she was acutely aware of the importance of taking care of her brain.

When I first met Kristy, she was positive and calm, but in looking at her brain scans, I saw something other than the placid, pleasant woman sitting before me.

After seeing the electrical activity of her brain, I realized that Kristy had excess beta brain wave activity—a sign of anxiety, stress, and the inability to relax. In other words, her brain was on fire—and not in a good way. Her nervous system appeared to be locked on high alert, and her neurons were firing too much in ways they didn't need to be, a condition that accelerates age-related cognitive decline.[8] This was akin to a doctor seeing a patient with 140/90 blood pressure—it was a giant red flag for Kristy's health.

After I saw her scan, I asked Kristy if she was stressed. That was when she told me the pipes had just burst in her newly renovated home, causing the entire downstairs to flood. She and her family had been living in a hotel for the last month, moving out just days after she finished costly renovations on her home. They were now stuck trying to assess the repairs and water damage.

Back to the car metaphor. This was a classic example of the exterior not matching the machine. Everything looked shiny and great on the outside, but inside, Kristy's engine was about to blow.

I couldn't fix her flooded home, take away her expensive repairs, or do anything else to remedy all the things that were stressing her out, but I could show her how to make specific lifestyle changes to calm her sympathetic nervous system, a division of the autonomic nervous system responsible for the body's fight-or-flight response.

After several months of practicing stress-reduction

techniques, the brain fog and fatigue Kristy described to me when we first met had resolved. She told me she went from feeling like she was just helplessly reacting to situations to taking control of her life, managing her stress more effectively while starting to take care of herself. This happened because Kristy finally made her brain health a priority. Today, Kristy lives a completely different life—and her brain is much healthier for it. Her life has dramatically improved, too, as increasing her cognitive health has also boosted her overall wellness and happiness.

The point of this story is that there's no way we can foresee every crisis or every potential stressor. But we can learn how to change our brain to help manage stress better. Perhaps more important, we can also learn to manage stress in support of healthier brain function.

KRISTEN'S TIP: Stress has a profound effect on brain structure and function, whether you feel it or show it. Learning ways to reduce your stress will boost your brain performance and make you sharper and healthier. Throughout this book, you'll learn how to better handle stress to calm your sympathetic nervous system for a healthier, happier brain and body.

Yes, You Can Change Your Brain at Your Age

A final word about changing the brain: You can do it no matter how old you are.

If you're in your twenties and assume you don't have to worry about cognitive decline, think again. The human brain doesn't fully mature until age twenty-five—some neuroscientists suspect the brain may even keep developing into our thirties. This means your current diet, sleep, exercise patterns, alcohol consumption, and overall lifestyle are impacting how your brain is developing.

In your thirties, your brain reaches full maturation—and then the process of natural brain aging begins. This is when we can start losing as many as eighty-five thousand neurons per day and quantifiable signs of cognitive decline can begin to appear. Taking care of your brain can slow the aging process and set you up for a healthier, happier, and smarter middle age.

When you reach forty, the volume of your brain begins to decline by an average of 5 percent every ten years. It's important to remember, however, that this is the average. You can slow age-related reductions in brain volume by adopting the new habits I'll outline throughout the book.

You can also start to experience lapses in short-term memory, reasoning, and verbal fluency in your forties.[9] At the same time, though, your brain's ability to regulate emotions and empathize with others is much more finely tuned.[10] Studies show concentration and sustained attention also peaks in your forties.[11]

In our fifties, studies show our overall knowledge base crests, and we're able to understand and learn new information better than we ever have or will.[12] This is one reason why researchers have found middle-aged people perform better on cognitive tests than they did in their younger years.[13]

While you may be your smartest in your fifties, your vocabulary skills don't peak until your sixties and early seventies.[14] Studies show pilots in their sixties fly planes better than their younger counterparts due to preservation of expert knowledge, even though it may take the older generation longer to scan the cockpit instruments.[15] And while our brain starts to shrink more rapidly in our seventies, researchers found those who stay physically fit and mentally active can be as happy and mentally healthy as twenty-year olds.[16] Brain scans also suggest people in their seventies have better emotional well-being than those in their twenties.[17]

If you're fortunate enough to reach your eighties or older, you have every reason to keep improving your cognitive health and function. If you take care of your brain, it will allow you to stay sharp so you can continue to engage with your friends and family, read books, enjoy movies, and pursue your hobbies. At the clinic, I even saw people in their eighties boost their cerebral circulation and improve their brain function. Remember that the brain can always change.

I have a saying I like to use with my clients: *No brain left behind.* We can make your brain better, no matter what your age is. Here's how to start right now.

10 Ways to Change Your Brain in 10 Minutes

1. **TAKE A BRISK WALK.** Research shows a short bout of exercise can increase cerebral blood flow, creativity, new idea generation, and overall executive function. If you've reached a mental block at work or need to prepare for a big meeting, do your brain *and* your career a favor by taking a fast walk around the office.

2. **EAT A SQUARE OF DARK CHOCOLATE.** Dark chocolate is rich in minerals and contains high levels of healthy plant compounds known as flavonoids, which can help to stave off free radicals and increase cerebral circulation and oxygen delivery. Some studies even show eating dark chocolate two hours before an event improves memory and reaction time.[18] Just be sure to stick to dark chocolate—neither milk nor white chocolate contains high flavonoid levels.

3. **SIT UP STRAIGHT.** Sitting up straight, with your shoulders back and neck long, can instantly increase blood flow to the brain. Studies also show sitting up straight can improve how others see you and increase self-confidence.[19]

4. **WRITE WITH YOUR NONDOMINANT HAND.** This little exercise asks your brain to go outside its everyday comfort zone, strengthening the neural connections and helping to spur neurogenesis. For some people, the mere act of writing by hand is a novelty for the brain, since they're accustomed to only texting or typing.

5. **SAVOR A BIG BOWL OF BLUEBERRIES.** If you're interested in growing new neurons, snack on a big bowl of blueberries. These berries are packed with flavonoids, polyphenols, and other healthy compounds that studies show can increase neurogenesis.

6. **LEARN A NEW WORD.** Expanding your vocabulary enhances your cognitive ability and overall intelligence while instantly adding new neurons to your hippocampus. Want a reminder to do this every day? Buy a page-a-day calendar that highlights a new daily definition, or download a dictionary app on your phone with this feature.

7. **VISUALIZE WAYS TO IMPROVE YOUR DAY.** Not only does this exercise calm the mind and lower stress, it also improves mood and can even optimize performance at work, in the gym, and in overall life. Professional athletes and CEOs often use the tactic before important events or make it part of their daily morning routine.

8. **CREATE TEN MINUTES OF WHITE SPACE.** Go into a room without phones and TV. No dings, beeps, chimes, chirps, news

feeds, broadcasts, or other distractions or demands—just you in a room with your eyes open or closed, enjoying ten minutes of time without any sources of stress. This exercise helps calm the sympathetic nervous system and can provide a greater sense of mental and emotional control for the rest of your day.

9. **SNIFF YOUR STRESS AWAY.** Using essential oils at home or in the office can help lower stress, calm the sympathetic nervous system, and alter brain wave activity to improve your cognitive function and mood.[20] Which scent is best? According to research, lavender is great for lowering stress, bergamot can help increase energy, and frankincense works to bring more oxygen to the brain.

10. **WRITE DOWN ONE THING YOU'RE GRATEFUL FOR.** Write down one thing you're grateful for on a sticky note and post it on your bathroom mirror, refrigerator door, office computer, or anywhere you'll see it throughout the day. Every time you do, the little reminder will help relax you, lower stress, and improve your mood.

2

BRAIN BASICS

My doctoral research was focused on understanding the role of the parkin gene and how mutations in this gene can lead to early-onset Parkinson's disease. To deepen my personal connection to the research, I started attending Parkinson's support groups to better understand the particular day-to-day struggles of those living with the disease. I was aware that the more I was connected to their unique physical, mental, and emotional challenges, the more my drive strengthened to not only address my research questions but also find ways I could be of service to this community. I was completely unprepared for the fact that, more than a decade later, I would find myself having to use this knowledge to help my own father.

When he passed away in 2017, it was a huge heartbreak. I loved my father dearly. If you knew him, you'd also have known how much of an inspiration he was, not just in my own life but in all of the lives that he touched. He was a proud marine, serving as a two-tour combat helicopter pilot with a marine corps squadron known as the "Ugly Angels" in the Vietnam War. He was also a captain with Pan American, flying 747s around the globe, and when not in the air, he loved serving his community for twenty-five years as an on-call firefighter. To me and many others, he was a true patriot and an American hero.

That's one reason why it was so difficult to watch my father, with his great strength, courage, and selflessness, begin to physically deteriorate to the point where he had trouble holding a pen, drinking from a glass, walking without a shuffle, or taking care of his two horses, Velvet and Zippy. At first, we didn't know it was Parkinson's. His tremors started two decades before he was diagnosed, which he and other members of our family passed off as simple age-related essential tremors. However, after I started working at the Amen Clinics, his symptoms progressed and became more difficult to ignore. I knew there was something neurologically wrong with him and that we needed to start doing things to support his nervous system.

After I realized my father had symptoms that were related to Parkinson's, I took my disbelief and heartache and turned it into action. I encouraged him to try nearly every protocol we were using with the NFL players to treat their cognitive problems, including hyperbaric oxygen therapy, nutritional supplements, and acupuncture. Given all that I had taught him over the years, he was already eating a brain-healthy diet, with a focus on organic, GMO-free, whole-food, and plant-based foods. I made a few modifications, swapping out white bread for wholegrain and whole milk for almond milk, and I encouraged him to eat at least two servings of seafood per week.

While he was completely open to making dietary upgrades, taking supplements, and regularly exercising, my father was a stubborn guy—he hated the idea of becoming a patient. That meant no hyperbaric oxygen chamber, special injections, fancy neurotesting, or getting poked and prodded by needles.

I'm still comforted by the fact that the changes my father did make helped improve his quality of life in small, meaningful ways. These changes allowed him to steady a fork with slightly more precision at a restaurant, which helped ease his embarrassment and motivated him to go out more often. He became a little steadier on his feet and better at holding his balance, allowing him to spend more time at the barn with his horses without worrying about falling (although he did lean on them for support)!

Whenever I think about my father, it breaks my heart that this physically strong man who'd flown thousands of people all over the world and soldiers in and out of war zones on rescue missions had become so weak. My hope through it all was that I was knowledgeable about the brain, and knew that however hopeless the situation seemed, there were lifestyle modifications and therapies that we could try. To empower him with tools that provided support to him in the years prior to his passing means the world to me.

I want you to understand the brain, too. It doesn't matter what you're struggling with, or whether you're struggling with anything at all. Your brain is an incredible organ and shouldn't be left behind because you don't yet understand the power you hold to be sharper, healthier, and happier.

Brain Basics 101: An Introduction to the Body's Most Complex Organ

The human brain weighs approximately three pounds on average, which varies from person to person, depending on height, weight, and sex. The average male brain is slightly larger than a woman's brain—1,274 cubic centimeters compared to 1,131 cubic centimeters, respectively.[1] While this makes men's brains about 10 percent bigger than women's, men are also on average physically larger than women, and their brain size reflects that difference. In this instance, bigger isn't smarter: studies show no difference in intellectual ability between the two sexes.[2]

While we share intellectual aptitude, there are subtle cerebral differences between the genders. Men appear to have increased connectivity from the front to the back of their brain, which can heighten their sense of perception and awareness of their surroundings.[3] Women, on the other hand, have increased connectivity from the left to the right sides of their brain, which may allow them to collate information more easily and draw more comprehensive conclusions. Some researchers, however, say these differences aren't congenital but biological by-products of how we are raised and socialized.[4]

The brain contains 100 billion neurons, or brain cells. Packed right in with all these neurons are just as many, if not more, glial cells, which act as support cells for the neurons, helping them coordinate activity across networks of neurons, transport chemicals, and clear metabolic buildup.

An individual neuron can connect to more than ten thousand other neurons through synapses, or junctions that allow cells to pass electrical messages, chemical signals, and other information. This level of connectivity keeps our brain insanely busy, capable of making more than 100 trillion connections. To put this number into perspective, 100 trillion is about one thousand times the number of stars in our galaxy.[5]

All these connections don't happen at a snail's pace, either: Neurons process information at a rapid-fire pace, transmitting by the second around one thousand nerve impulses, or signals, across synapses.[6] Some impulses travel from body to brain or vice versa very quickly, at speeds of up to 268 miles per hour—faster than a Formula One racecar. Other information can lag along at much slower rates, or about one mile per hour.[7]

No matter the speed, the frequency of this neuronal activity is incredible, causing your brain to generate real electricity. In fact, your brain can produce enough electrical activity to power a low-wattage lightbulb. Several years ago, one science writer even figured out our brain could generate enough electricity to fully charge an iPhone 5C in about seventy hours.[8]

Neurons send signals to other neurons in order to perform a specific function, like processing what you see with your eyes or recalling the name of a colleague or friend. This chain of intercommunication between neurons is known as a neural network or pathway. When neurons communicate over and over again in the same neural network, that network grows stronger.

But neural networks aren't like interstate highways, offering riders the same, fixed route from point A to point B. Quite the opposite, our neural networks can change paths often, capable of being redirected or even destroyed by the lifestyle habits we sustain over time. We can also create new neural pathways by learning novel information, adopting certain habits, and challenging the brain in other ways.

When it comes to mental-storage capacity, the brain outperforms your iPhone and even your desktop computer. The human brain can hold the equivalent of 2.5 million gigabytes of digital memory[9]—compare that to even the newest smartphone, which can store only 512 gigabytes. To look at it another way, if our brain were capable of recording TV, it could hold around 300 million hours of shows, enough to keep your TV continuously running for more than three hundred years, according to *Scientific American.*[10]

Some scientists like to compare the brain to a computer, but the actual composition of the brain is nothing like the inside of a PC. Instead, our brain is made up of 75 percent water by weight and 60 percent fat by composition. This means our brain needs to be well hydrated, which allows for good blood flow, along with being provided essential fatty acids (EFAs) from our diet. Our body can't produce EFAs but has to get them from what we eat, making our diet critical to overall brain health (we'll delve deeper into this in chapter three).

Our brain also needs a continual supply of glucose, or sugar, in order to operate. Unlike our muscles and liver, our brain can't store glucose but must rely on proper blood flow to get the sugar needed for neurons to fire and function efficiently. Our brain also requires lots of vitamins, minerals, electrolytes, and other nutrients from our diet.

How This TV Host Started Tuning In to the Reality Show of Brain Health

Most people know Mark Steines from TV. The Emmy-winning broadcast journalist and sports anchor has been seen by millions on the small screen for more than three decades, most notably as the host of *Entertainment Tonight* and *Home and Family*.

What many don't know about Mark, however, is that before he launched his career in TV, he was a formidable football player, attending the University of Northern Iowa on a full athletic scholarship. He played for eleven years as a fullback, with dreams of going to the NFL, but his athletic career ended when he tore his ACL his senior year. The injury prompted him to turn to TV, but he never lost his love of sport and still remains connected to football today, hosting the annual Rose Parade and interviewing pro players.

I have had a wonderful friendship with Mark since

2005. When my colleagues and I started our research with NFL athletes, he avidly followed our work. In 2013, he invited me to join him on *Home and Family* to discuss our findings. When I talked openly in front of his audience about the real effects that hard hits have on the brain, he was both amazed and concerned.

Despite being immersed in the sport for decades, Mark knew little about what football could do to the brain—typical of almost everyone at the time when we were conducting our research. I was the first person to open his eyes to the real consequences that concussions have and the serious cognitive damage that can occur years later simply from participating in one of America's most cherished pastimes.

Before I met Mark, he had the attitude many former players did at the time. They saw getting their "bell rung," as they put it, as a natural corollary, even a rite of passage, of the sport that resembled combat warfare more often than it did an athletic endeavor. At the time he played ball, if you took a hard hit and had to sit out, you were perceived as weak. Instead, you shook it off with a smile and headed back into the game, without medical attention or even a moment's time out.

Through our friendship, Mark began to understand that the brain wasn't this abstract object inside his head, capable of taking hit after hit without sustaining any damage. The brain was no different from any other part of his body—knee, groin, neck—except that it was a lot more precious, and if you injured it, it wouldn't be as easy to heal as a torn ACL, strained groin, or even ruptured disc.

As Mark learned more about the brain, his concern evolved past how it would impact his own life, as he grew more concerned about his two young sons, who were playing flag football at the time. Up until the time our research was published, he had dreamed that one of his sons might be able to take his athletic career further than his father had. But now, he was reconsidering whether he should let them even play flag football in the first place. Fortunately, both of his sons were more interested in pursuing artistic endeavors and eventually left the sport (one is now a musician, the other is a sound technician).

Today, Mark thinks differently about his brain than he did while playing football. His mind used to be something he had to fight against when it told him something he didn't want to hear—*you can't run as fast as the competition, you won't play pro ball if you sit out, you need to rest*. Instead, he'd tell himself, *Get out of your head, don't overthink it.* Like many players, he was conditioned to put his head down and go, ignoring what was going on inside his mind.

Now, though, Mark says he pays attention to his brain because he's realized how critical cognitive function is to his present health and future happiness. His mind is the one show he knows he'll host for the rest of his life and is what will keep him able to speak fluently on TV, retain his memory, and have meaningful relationships with his family for as long as he can.

KRISTEN'S TIP: If you're a parent or grandparent of children involved in impact sports, learning more about the brain can help you make better decisions about their athletic future. Take the information you learn in this book to talk with your spouse and/or child's coach about what's best for his or her cognitive health.

Where Higher Thinking Happens and the Truth About Your Two Hemispheres

Most of us know our brain has two types of tissue: gray matter and white matter. But what do these tissues actually do?

Gray matter, named for its pale, pinkish-gray color, contains most of the brain's neurons and the areas that process information and make you capable of thinking, reasoning, and remembering.

White matter, on the other hand, is made up mostly of nerve fibers that allow for efficient communication between neurons in the brain. White matter gets its name from the fatty, pearl-colored material called myelin that surrounds axons. If you stretched out the myelin-insulated axons in the brain of an average twenty-year-old, you'd have about one hundred thousand miles of nerve fibers.[11] That's more than four times the Earth's circumference.

In addition to gray and white matter, our brain has three primary parts: the cerebrum, cerebellum, and brain stem. The cerebrum is the largest part of the brain, taking up about 80 percent of your brain volume.[12] The cerebrum sits in the forefront of the brain and is responsible for higher cognitive function, including learning, thinking, problem solving, language skills, and memory. The cerebrum also interprets sensory information we get from our eyes, ears, skin, nose, and mouth, and controls many of our feelings and emotions.

Your brain's cerebrum is divided into two sides, or hemispheres: the right hemisphere and the left hemisphere. Your brain's left hemisphere controls the right side of your body and vice versa. The two hemispheres connect with each other through a thick band of some 200 million nerve fibers called the corpus callosum.

Most of what we know about the brain's two hemispheres comes from studies conducted on patients who have had their corpus callosum severed. Doctors performed the procedure for years in order to prevent epileptic seizures from moving from one hemisphere to the other. The procedure, no longer used, meant cutting the communication cable between the two sides of the brain, which created a condition called split-brain syndrome. Split-brain patients can act normally in many ways, but will then fail to recognize certain objects or recall common words. They also can't learn new skills that involve independent movement of each hand, like playing the piano.

While some believe righthanded people are controlled by the left side of their brain and lefthanded people are controlled by the right side of their brain, there is no scientific evidence to support this.[13] Right- and left-handed people use both sides of their brain. However, left-handed and right-handed people can use the brain's hemispheres in different ways.[14]

The left side of the brain is responsible primarily for language, comprehension, math, and writing skills. Approximately 97 percent of people derive their speech ability from the left hemisphere, with little to no contribution from the right side.[15] The left hemisphere also contains more neurons than the right.[16]

The right hemisphere is where we derive our spatial orientation, visual and artistic abilities, face recognition, and musicality. The right side also arbitrates our emotions and nonverbal communications.[17]

Each side of our brain has four lobes. The frontal lobe, the largest of the four located in the upper front of your brain, is headquarters for the brain's executive function, or higher thinking. The frontal lobe also controls voluntary physical activities, or what you choose to do with your body and limbs, like walk across the room or reach up with your arm for an item.

The parietal lobe, located behind your frontal lobe in the upper rear of your brain, translates sensory information like taste, temperature, touch, pressure, and pain. This lobe also aids in spatial recognition, along with skills like reading and math.

The temporal lobe, located above your ears below the frontal and parietal lobes, helps process memories and sound. Finally, the occipital lobe, the smallest of the four situated in the back of your brain, coordinates aspects of your vision.

Some researchers say our brain has a fifth lobe, the limbic lobe. This area of the brain exists—it's a horseshoe-shaped structure located deep inside the brain that helps regulate instinctual behavior, emotions, and memories. The lobe is part of what scientists call the limbic system, which controls our emotional response to events and conditions, along with our memory recall and feelings related to hunger, fullness, sexual arousal, and other sensations.

In addition to the cerebrum, the other primary parts of the brain are the cerebellum and brain stem. The cerebellum, located under the cerebrum in the back of the brain, is an intricate ball of neurons that helps control movement and balance, eye movements, and the fine-motor skills we learn over time, like riding a bike or playing a musical instrument. While the cerebellum represents only 10 percent of the brain's total volume, it contains more than 50 percent of the brain's neurons, making it very valuable neural real estate. There are, however, a few reported cases of people who have been able to survive without a cerebellum.[18]

The brain stem sits under the cerebellum and extends down the neck, connecting the brain to the spinal cord. The brain stem is one of the oldest parts of the brain and controls the body's automatic functions, including breathing, heart rate, temperature, and digestion. The brain stem also regulates involuntary muscle movements like the beating of your heart and the expanding and relaxing of your lungs, while processing millions of messages between your brain and body.

Brain Biology: Six Areas to Get You Started in Understanding Neuroscience

For the purpose of this book, you don't need a degree in neuroscience. Really, you just need to begin by understanding the most important structures: the cerebral cortex and those that make up our limbic system.

THALAMUS: Located in the middle of the brain, the thalamus is often called the central hub, or relay station, for the rest of the cerebrum. Here, the brain processes most external sensory information, with the exception of smell, and sends that info to other areas of the brain. The size of only two walnuts, the thalamus also helps control pain sensation, attention, alertness, and some types of cognitive thinking.

HYPOTHALAMUS: This almond-sized structure sits below the thalamus and connects the nervous system to the endocrine system, which controls hormones. The hypothalamus helps release hormones from the brain's pituitary gland, regulating sleep, hunger, thirst, body weight, and temperature, among other conditions.

HIPPOCAMPUS: This is the most critical area of the brain for memory, new brain-cell growth, and cognitive disease prevention. Located deep inside the brain below the thalamus and hypothalamus, the hippocampus is responsible for storing and recalling memories, learning information, and coordinating spatial navigation. Loss of brain volume in this cerebral structure contributes to cognitive decline.

AMYGDALA: Located near the hippocampus, the amygdala is one of the brain's most fascinating structures, responsible for helping control our fear response, emotions, and sense of pleasure. The fear response triggered by the amygdala can be intense and visceral—one reason some refer to it as the brain's fear center. Sometimes, the amygdala can respond with fear before other areas of our rational brain have time to "think" about it, subsequently shutting down neural pathways that allow for good decision-making.[19] Called an "amygdala hijack," the condition can cause people to overreact.

Size matters when it comes to the amygdala: The bigger it is, the more aggressive you're likely to be. Studies show those with psychopathic tendencies tend to have larger and more active amygdala.[20] Some people have even undergone surgery to remove the amygdala in order to limit aggression and curtail fear and anxiety.

CINGULATE GYRUS: This curved belt-like structure in the middle of the brain wraps around your corpus callosum—the neuron mass that separates the left and right hemispheres. The cingulate gyrus has two parts—the anterior cingulate gyrus and the posterior cingulate gyrus—and is generally seen as the interface between the areas in the brain involved in high-level cognitive function and the emotional center in our limbic system. As a whole, the cingulate gyrus helps regulate our emotions, motivation, decision-making, memory, learning, and some autonomic physical functions.

PREFRONTAL CORTEX: While not part of the brain's limbic system, the prefrontal cortex is an important area to know. The prefrontal cortex is the front portion of the brain's overall cortex—a thin layer of folded gray matter that covers the cerebrum and gives the brain its wrinkled, walnut-like look. The prefrontal cortex plays a pivotal role in many executive functions like focus, impulse control, planning, reasoning, decision making, and anticipated behavioral adjustments.

Secrets of the Conscious, Subconscious, and Unconscious Mind

If you're surprised to find a section on the conscious mind in a brain book, you're not alone. For most of the twentieth century, researchers believed consciousness had little to do with neuroscience. It wasn't until the 1990s with advances in brain imaging that neuroscientists began to understand the brain profoundly affects consciousness and vice versa.[21] This realization was further fueled by studies showing coma patients still demonstrated brain activity in response to verbal cues, suggesting the conscious mind is still at work even when the brain is essentially unconscious.[22]

The conscious mind is the seat of all thoughts, emotions, and memories we're aware of. It's where we do all our rational reasoning and what gives us our unique ability to have free will—it is, essentially, what makes humans different from other life on Earth.

Most people wrongly assume all our thoughts and ideas originate in the conscious mind. In reality, the conscious mind is responsible for only 10 percent of the fifty thousand to seventy thousand thoughts we generate on a daily basis. The majority of our thinking, according to Austrian neurologist and psychoanalyst Sigmund Freud, comes from the subconscious mind, which makes up 50 to 60 percent of our daily thoughts. The unconscious mind, on the other hand, is responsible for the remaining 30 to 40 percent of our thinking.

The subconscious, sometimes called the preconscious, is considered a subset of the conscious mind and is where we store memories, habits, and behaviors we don't need at the moment.[23] We can tap our subconscious when we want to recall a random memory like a phone number, what we ate for lunch yesterday, or what time a work meeting will take place the next day.

The unconscious mind, on the other hand, is where deep memories we can't recall at will are stored, along with enduring emotions, habits, and behaviors, many of them programmed in our brain since early childhood.[24]

What do the conscious, subconscious, and unconscious have to do with cognitive health? Simply put, every thought, action, and behavior influences brain power and performance. With the right guidance, many of us can learn to better control our conscious thoughts.

Elizabeth's Story

HOW LEARNING ABOUT BRAIN WAVES AND THE CONSCIOUS MIND TRANSFORMED HER LIFE

I had the pleasure of meeting Elizabeth,* a board-certified hypnotherapist, at a physical therapy center several years ago.

Her job essentially entails guiding the patient into a trance-like state using techniques to quiet their conscious mind. This allows them to connect to their subconscious mind and access deeper states of awareness to better manage reoccurring issues like chronic pain, anxiety, addiction, depression, fear, and phobias.

* Some client names changed to protect patient privacy.

To put this in perspective, we can enter a trance-like state when we daydream, get immersed in a good book, drive on autopilot, cook, meditate, or exercise. Hypnotherapy allows practitioners to access that deeper brain wave state and help guide patients to overcome problems locked in their subconscious mind.

While Elizabeth had been practicing for a number of years, her knowledge of brain biology from an imaging perspective was limited. I taught Elizabeth about brain waves, or the bioelectric oscillations created when the brain's neurons communicate with one another. These waves vary in frequency and amplitude—from high frequency and low amplitude to low frequency and high amplitude—and can be tuned up or down, if you know how to do so. Tuning up brain wave frequency can create a state of heightened awareness, but also heightened anxiety, energy, and arousal. Tuning down brain wave frequency, on the other hand, can reduce stress and unlock a deeper state of relaxation.

I thought if Elizabeth understood the science behind brain wave oscillation, she might be able to shift her patients more easily into slower frequencies that would allow her to access their subconscious and unconscious mind. It might also help her identify which patients would be more responsive to treatment and how to work with people who didn't respond to hypnosis.

My assumptions were right: after several months together, Elizabeth told me her new knowledge enabled her to transform her practice, helping her make breakthroughs with patients in ways she never thought possible.

Our work together also led to another important result: Elizabeth was able to learn more about her own brain and how powerful her thoughts could be on her physical, mental, and cognitive health. Through our work, she became aware that every single thought creates a chemical reaction in the brain that can impact the body. She began paying closer attention to what bubbled up from her subconscious and unconscious into her conscious mind, making sure the thoughts she could control were positive. You'll learn more about the power of thoughts in chapter eight (see page 163).

While these ideas may sound newfangled, there is nothing far-fetched about the brain-body connection and its potential significance on our overall health and happiness. As Elizabeth learned, gaining even a little more insight into the brain's biology and how cognitive thinking can affect our body can have a profound and lasting effect on our overall quality of life.

KRISTEN'S TIP: Understanding your own brain and how it functions can help you better control negative thoughts and deepen your understanding of the mind-body connection.

Intellect: What It Really Means to Be Smart

You can't read or write a brain book without having at least one discussion about how much intelligence is rooted in neuroscience and, more important, the ways in which intellectual capacity can be enhanced.

Intelligence, quite simply, is the general mental ability to reason, solve problems, and learn new information. It involves perception, memory, attention, language, and planning. What isn't so simple is where intellect actually originates in the brain. Today, most researchers agree that multiple areas of the brain work together to create and maintain intelligence.[25] Neurobiological studies show genes help determine some but not all of our smarts. Genes can also influence the size and efficiency of the cognitive areas responsible for higher-level thinking.[26]

The genetic factor doesn't mean intelligence is crystallized, or incapable of changing over time. Remember the brain is incredibly plastic, capable of being molded, both to our benefit and detriment. In terms of intelligence, research shows we can change our brain to boost intellect in a number of ways. Primary among them, perhaps unsurprisingly, is by learning new information and skills, which helps to strengthen neuronal communcation and rewires parts of the brain responsible for cognitive thinking.[27]

Another significant way to improve intelligence is by changing how we eat, exercise, sleep, and handle stress. According to research, small behavioral changes can make your brain more efficient in as little as two weeks' time and increase cognitive intelligence over time.[28]

Mental attitude and the power of your thoughts also play a big role in overall intellect. For example, one study found students who were taught they could change their intelligence improved their grades more than those who never learned that intellect is malleable.[29] What this means is that knowing and believing you can change your brain will help you increase your cognitive function.

Fascinating Facts and Shocking Truths About Our Brain

I may be biased toward the brain, since I've spent my life studying this incredible organ. But I can guarantee you there are

What About IQ?

IQ, short for "intelligence quotient," was created by psychologists in the early 1900s as a way to assess and rank academic progress. IQ scores are determined from a series of cognitive tests. The average IQ score is 100: If you score below that, you're considered less intelligent, while those who score above 140 are said to be geniuses. (Both Stephen Hawking and Albert Einstein are estimated to have IQ scores of 160.)

Many neuroscientists and modern-day psychologists are skeptical of IQ, saying it measures only academic ability, not innate intellect. To that end, studies show people can increase their IQ simply by changing their level of schooling, family environment, work environment, and even style of parenting.[30]

fascinating facts and shocking fictions about the brain that will surprise and delight even the most neurophobic among us.

First, it's fiction that we use only 10 percent of our brain. Instead, we use 100 percent of the brain, even when we're resting and sleeping. In fact, our brains are quite active while we sleep, doing important things like cleaning up waste created during the day.

Another myth: humans have the biggest brain on Earth. That distinction belongs instead to the sperm whale (the same creature of Melville's *Moby Dick*), which has a brain five times the size of ours. The sperm whale's big brain makes it smarter than most mammals, too. Studies show sperm whales are excellent communicators, and while they're too large to be the subject of research projects, studies on marine mammals in the same family show they can self-recognize in front of a mirror and even be trained to find underwater mines and military personnel lost at sea.[31]

If you're a parent, you've likely heard playing classical music to your child will make him or her the smartest kindergartener in the class. Well, I have unfortunate news for all the Beethoven babies: listening to classical music won't boost an infant's IQ, despite a plethora of playlists and DVDs that claim otherwise.[32]

Our brain also can't feel pain, even though we process sensory information for pain in the brain. Doctors can even perform brain surgery on a patient without anesthesia and without causing any discomfort. What about headaches? Despite what may feel like a pounding pulse in your brain, headaches may be caused by strained muscles, sinus issues, narrowing blood vessels, and other issues that don't originate in the actual brain.[33]

Speaking of headaches, you know the brain freeze you get when you drink or eat something cold quickly? The so-called ice-cream headache isn't something to worry about: It doesn't cause damage to brain cells to scarf down an entire pint of chocolate-chip cookie dough (what this does to your blood sugar and insulin is another story, however). Brain freeze is caused by a sudden contraction of blood vessels around the brain.[34] According to scientists, a brain freeze is actually a good thing, because it tells you to slow down when consuming cold things quickly, ultimately helping to protect the brain's internal temperature.[35]

Another beneficial brain adaptation is forgetfulness. Our brain has built-in mechanisms for memory loss so that we don't waste precious storage space recalling irrelevant details that can get in the way of the information we really need to know.[36] The next time you forget someone's name or where you put your purse, don't freak out—it's probably just part of your brain's game for life.

While there is no such thing as left- or right-brained people, the shape of our brain does influence our personality, according to studies. Research shows people who are more open-minded, curious, and creative have a thinner cortex—the wrinkled outer layer of the cerebrum—and a larger cortical area with an increase in folding, allowing the brain to hold more neurons. Those with neurotic tendencies, on the other hand, have a thicker cortex, a smaller cortical area, and less advantageous brain folding.[37]

Interestingly, studies show introverts have larger, thicker gray matter in the prefrontal cortex than extroverts do. Since the prefrontal cortex is associated with abstract thought, scientists think the structural difference may be due to the fact that introverts spend more time in abstract thought than socializing with others, causing the brain to change. That's just more proof of how plastic our brain can be.[38]

Seven Common Conditions That Can Affect the Brain

The more you know about the conditions that affect cognitive health, the less fear these ailments can wield over you. Throughout this book, you'll learn how to proactively prevent and combat many of these conditions for a healthier, happier brain.

1. **ALZHEIMER'S DISEASE**: An estimated 5.8 million Americans have Alzheimer's, affecting one in ten people age sixty-five or older, making it the most common neurodegenerative disorder in the country.[39] While research is evolving, the disease is believed to be caused by abnormal deposits called amyloid plaque in the brain, along with tangled bundles of fibers called tau. Signs of Alzheimer's can manifest in brain scans decades before patients show symptoms, which can include memory loss, difficulty understanding and

completing familiar tasks, decreased judgment, and other behavioral and social problems.

2. **PARKINSON'S DISEASE:** Parkinson's is the second most common neurodegenerative disorder in the United States, affecting approximately 1.5 million people. The disease can cause people to suffer muscle rigidity, tremors, and overall difficulty with physical movement. Researchers aren't sure what causes Parkinson's, although both genetic changes and environmental factors like exposure to chemicals and head trauma can play a role. In addition to those listed above, symptoms include problems with posture, handwriting, and speech.

3. **DEMENTIA:** Dementia isn't a disease itself, but a term used to describe a broad condition marked by a significant decline in cognitive function that interferes with memory, rational thinking, and social aptitude. Alzheimer's is the most common form of dementia, contributing to 60–70 percent of cases, while those with Parkinson's are at risk for dementia as the disease progresses. It's estimated that one out of every six women and one of every ten men over the age of fifty-five will develop dementia.

4. **MILD TRAUMATIC BRAIN INJURY (CONCUSSION):** Over the past few decades, mild traumatic brain injuries have become a major talking point among athletes due to the increasing incidence of concussions in contact sports. A mild traumatic brain injury occurs when someone sustains a blow to the head and doesn't lose consciousness for more than thirty minutes—if they do, it's considered a traumatic brain injury.[40] An estimated 1.6 to 3.8 million concussions occur annually in the U.S. during sporting events and other recreational activities.[41] Symptoms, often subtle, include mem-

ory loss, fatigue, headaches, visual impairment, and mood changes. Sustaining too many repetitive subconcussive impacts or concussions over time can lead to chronic traumatic encephalopathy, or CTE, a degenerative brain disease found mostly in football players and military vets.

5. **ANXIETY:** Anxiety is the most common mental health issue in the United States, affecting 40 million Americans age eighteen and older on an annual basis. What does mental health have to do with the brain? In the instance of anxiety, researchers say the condition originates in the brain with overstimulation of certain neural pathways and areas such as the amygdala. Mental symptoms can include feeling worry, nervousness, restlessness, and unease, along with fatigue, irritability, muscle tension, and sleep problems.

6. **DEPRESSION:** Similar to anxiety, depression arises from problems in brain biology. Chemical imbalances can cause depression, but it's often not as cut-and-dry as changes in neurotransmitter levels. There are millions of different chemical reactions that occur in the brain, and chemistry alone doesn't determine depression—changes in areas like the hippocampus can also play a pivotal role. In addition, depression can be influenced by genes, work, sleep patterns, prescription drugs, and other factors. According to the National Institutes of Health, more than 17 million people have experienced at least one major depressive episode.[42] The Centers for Disease Control and Prevention also estimates around 8 percent of the total U.S. population is depressed during any given two-week period.[43]

7. **STROKE:** Similar to traumatic brain injury, strokes can cause significant neurodegenerative effects. A stroke occurs when blood flow is blocked to the brain, cutting critical oxygen

and nutrient supply to cells. This, in turn, can cause neurons to die quickly, triggering a cascade of physical and cognitive problems. Immediate symptoms include sudden paralysis (usually isolated to one side of the body), difficulties speaking and understanding, vision impairment, and overall muscle weakness. Strokes can lead to memory loss, impaired cognitive thinking, and permanent brain damage. Every year, nearly eight hundred thousand Americans experience a stroke, with someone in the United States suffering from a stroke every forty seconds.[44]

3

THE BETTER BRAIN DIET

After we began our study with NFL athletes at the Amen Clinics, we had recruited just fifteen players when we realized the damage to their brains was more extensive than we imagined. Since most were overweight or obese, our first priority was to encourage them to lose weight, since excess body fat severely impacts brain function.

Nevertheless, I was still surprised when Dr. Daniel Amen, the founder of the clinic, asked me to lead a weight-loss group for the players. I thought to myself, *I'm a neuroscientist, not fitness expert Jillian Michaels. These are professional athletes! They already know how to eat and train to get into competition-level shape.*

Since a majority of the players lived in different areas of the country, I held bimonthly sessions online when everyone could call in and participate. I created PowerPoint presentations, teaching them how to eat to fuel their brains. To my surprise they came prepared with questions, wanting to gain a deeper understanding of how proper food choices can positively impact brain health. The group lasted more than a year, but it didn't take long for it to turn into a close-knit community, as they shared stories of overeating, cravings, and control around food. It also didn't take long before I became a trusted resource and this is when they gave me the nickname "Coach K."

During our sessions, I taught them the best foods for the brain, the dangers of artificial sweeteners, how to read food labels, why to eat organic, how to eat more omega-3s, the right way to follow the Mediterranean diet, and how and why to eat low-glycemic. I created a variety of meal plans to work with, including a Mediterranean option, a low-glycemic option, and one using foods from a national grocery-store chain that specializes in organic, non-GMO, and premade foods for those who were not fans of cooking. I even showed them a picture of my father's shopping cart, which was always brimming with brain-healthy foods. If a seventy-year-old man who struggles with tremors could find the inspiration to shop and eat for his brain health, these guys could, too. At the end of every class, I included a recipe that combined several brain-friendly foods.

While I enjoyed teaching undergraduate courses at UCLA, I hadn't truly realized how much I would love teaching the players to take care of their brains through making better lifestyle choices. It certainly helped that the players were very coachable—they had received instruction their entire lives and responded with a proactive attitude, which was why they were so successful in making changes. They were excited to take direction and do the necessary work to overhaul their nutritional choices.

Over time, the group turned into a family of athletes with the same common goal: to get brain fit. The group had been so successful that I began running a similar session for the patients at the clinic. Many of the players wanted to continue to attend, and some did for years afterward, allowing me to remain a close and active part of their cognitive-health journey.

Throughout our sessions, brain health was always the priority—weight loss was the beneficial side effect. Everyone wants to lose weight, so the sell was easy. Many also couldn't believe they could drop fat simply by following a food plan designed to boost their cognitive health.

I'm proud to say that everyone who wanted to lose weight did. The weight loss was on a spectrum: some lost thirty pounds, while others lost up to eighty pounds. At the end of the day, it wasn't just about weight loss; it was more about creating and reinforcing brain health habits that would be with them for a lifetime.

The Better Brain Diet, outlined here, is a seven-step process that incorporates the basic principles that I covered during the NFL weight-loss group. The plan encourages making small, incremental changes to your daily diet for the sake of your brain. The primary goal is to get your brain fit—however, if you want to lose weight, you can and will.

I suggest you read each step carefully, even if you believe you already adhere to it, as there are specific recommendations you may not know about. In "Putting It All Together" on page 73, I specify how many servings of each food group to aim to eat every day, where alcohol fits in, and how you can modify the diet to make it your own.

Step One: The Simple Food Swap That Will Save Your Brain

In today's world of easy convenience and packaged everything, Americans consume nearly 60 percent of their daily calories from processed foods, or ready-to-eat, packaged products made mostly from synthetic ingredients manufactured in labs, not grown on farms.[1] Processed foods include potato chips, crackers, breakfast cereals, frozen dinners, soda, diet soda, cookies, candy, ketchup, many salad dressings, pasta, bread, fruit yogurts, deli meats, and the vast majority of items found on store shelves.

What's so terrible about processed foods? In a nutshell, everything. They're high in calories, sugar, unhealthy fat, useless carbs, and harmful chemicals. At the same time, they're devoid of the nutrients the brain needs not only to function optimally but also to survive.

In working with clients, I've found many incorrectly assume their sugar levels are relatively low—they tell me they don't eat candy or drink soda, and only occasionally indulge in dessert. But sugar is in nearly everything we eat, not just typical items like soda, juice, cookies, jams, cakes, and candy but also smoothies, protein bars, yogurt, breakfast cereal, bread, crackers, ketchup, salad dressing, store-bought sauces, sports drinks, and coffee drinks. Even foods that are organic, vegan, low-fat, or gluten-free can be high in sugar. Sugar is so universal that Americans, on average, consume around seventeen teaspoons of sugar per day—that's approximately ten teaspoons more per day than our dietary guidelines recommend.[2]

Not only are processed foods sugar bombs, they're also packed with chemical additives like preservatives, emulsifiers, synthetic colors, hydrogenated fats, artificial sweeteners, fake flavorings, MSG, and known carcinogens like acrylamides. These additives help make foods taste better, appear more aesthetically appealing, or last for months, if not years, on store shelves.

But food additives don't belong in food. Studies show these chemicals can increase the risk of cancer, heart disease, diabetes, and nearly every other chronic illness, along with the overall incidence of premature death.[3] In the brain, the chemicals can impede memory and concentration, restrict blood flow to the brain, and increase the risk of neurocognitive decline and disease.

The only good thing about processed foods is that they're simple to avoid: just eat whole foods. Whole foods are what you'd eat if you lived before the onset of industrial food manufacturing—foods that come straight out of the ground or from a farm, with little to no additions or alterations.

While the term "whole food" has become cliché in recent years, it's important to understand what they are and why they're imperative to cognitive health. In short, whole foods combine the type of complex carbs, fats, protein, fiber, vitamins, minerals, antioxidants, and other nutrients your brain and body need to function properly. Every time you eat a whole food, it's like swallowing the strongest multivitamin. They're also free of added sugars and chemical additives.

If you do nothing to your diet other than swap out processed foods for whole ones, you will increase your cerebral circulation, grow new neurons, and reduce inflammation, not to mention also incur a whole host of other health benefits.

Step Two: Eat More Fat, But Only the Right Kinds

Our brain is 60 percent fat, which means dietary fat plays an essential role in cognitive function. Fats, or lipids, are needed to form neuronal membranes and for the cells to function properly. Fat is also what makes up the brain's myelin sheaths, which surround nerve fibers and enable neurons to transmit messages quickly and efficiently. Consuming fat is critical to optimal cerebral circulation, neurogenesis, and nearly every structure and function responsible for higher thinking. Without enough dietary fat, you also increase your risk for neurodegenerative decline and diseases like Alzheimer's and Parkinson's.

The One Fat We All Need to Eat More Of

Not all fats, however, are beneficial to the brain. The type of fat the brain needs the most is also the one we consume the least: essential fatty acids (EFAs). EFAs, in particular the ones found primarily in fish and seafood, are critical to the brain's ability to function optimally.[4]

Our body can't manufacture EFAs on its own—we have to get them from our food or supplements. While the Institute of Medicine's Food and Nutrition Board has not instituted a recommendation for omega-3 EPA or DHA intake,[5] the American Heart Association recommends people consume at least two 3.5-ounce servings of fish per week to increase EPA and DHA.[6] These guidelines are essential as authorities estimate upward of 90 percent of Americans don't get enough marine omega-3s through seafood.[7]

There are three types of omega-3s: alpha-linolenic acid (ALA), found in nuts, canola oil, flaxseed, and other plant foods; eicosapentaenoic acid (EPA); and docosahexaenoic acid (DHA), found in fish and seafood. While consuming more omega-3s of any kind can boost brain function, the most important EFA is DHA.

DHA comprises 90 percent of the fatty acids in the brain. This EFA is imperative to nearly every cognitive function, including neuron survival and growth, neuroplasticity, synaptic transmission, cerebral circulation, and cell-membrane integrity, which will support your memory, concentration, problem-solving, and information-processing abilities.

ALA is more abundant in the average American diet, since it's found in many plants we regularly eat like nuts and beans. Our body can convert some ALA we consume into DHA, but only about 15 percent into both DHA and EPA.[8] To improve brain health and function, you need to consume foods with DHA.

Remember, your body can't manufacture significant amounts of this fat on its own. Here's how to eat more of this essential EFA for a better brain:

1. **EMBRACE THE OCEAN.** The foods highest in DHA are salmon, tuna, trout, mussels, herring, mackerel, sardines, and other seafood. In general, oily cold-water fish have more overall omega-3s than bass, tilapia, and cod. Better still, you don't need more than one or two servings per week to improve DHA intake.

2. **NOT ALL FISH IS "FISHY."** If you don't like the "fishy" taste of seafood, try flounder, haddock, catfish, trout, and arctic char. Fish, especially some varieties like cod, are like chicken: They soak up whatever seasoning or sauce you put on them. Vary your preparation—think fish tacos, fish burgers, baked fish sticks with an oat crust, and soups made from shellfish.

3. **CHOOSE YOUR SEAFOOD WISELY.** Both wild-caught and farmed species can contain toxins like polychlorinated biphenyls (PCBs) and mercury that can damage the brain, and you should do everything you can to reduce exposure to these potentially harmful chemicals. Before you buy or order at a restaurant, check the fish's safety profile on the Monterey Bay Aquarium's Seafood Watch app or its website. Both rank seafood based not only on health but also on species sustainability.

4. **LEARN TO LOVE THE TWO "S" PLANTS.** Short of supplements, seafood isn't your only option for DHA. Algae, particularly seaweed and spirulina—a type of blue-green algae—also contains this hard-to-get omega-3, albeit in lower levels. In addition to containing DHA, both also have a high concentration of micronutrients. We'll learn more about spirulina and how to supplement with it in chapter five.

5. **OPTIMIZE YOUR ALA.** While seafood is highest in DHA, we can covert some ALA into DHA and EPA. To give your body the best benefit for conversion, choose foods high in ALA like chia seeds, hemp seeds, flax seeds, walnuts, edamame, and kidney beans. Some plant oils, like flaxseed, walnut, hemp, and chia, are also rich in ALA.

6. **BUYER BEWARE IF YOU FORTIFY.** Given America's dismal consumption of omega-3s, many products on store shelves are fortified with DHA and EPA, including breakfast cereals, orange juice, energy bars, salad dressings, and even baked goods. But many of these foods are highly processed and contain more sugar, refined carbs, and/or chemicals than they do DHA.

The Truth About Saturated Fat and Cholesterol: What to Eat and What to Avoid

Over the past decade, debate has erupted in the dietary world over whether saturated fat and dietary cholesterol are the nutritional villains many have made them out to be. Your brain, like the rest of your body, needs saturated fat and cholesterol to function optimally. Saturated fat in particular is key to making cell membranes, while cholesterol, among other physical functions, helps produce hormones that play a critical role in cognitive health.

But those who say we should eat more bacon, steak, eggs, and cheese are mistaken about what's best for brain health. Nearly every large, long-term study shows consuming too much saturated fat and unhealthy LDL cholesterol can have disastrous effects on the brain, causing inflammation, memory impairment, and mood dysfunction, and increasing the risk of Alzheimer's and other diseases.[9] (For more on LDL cholesterol, see page 202 in chapter ten.)

Because we still need saturated fat and cholesterol for proper brain health, I recommend coconut and coconut oil. Both are high in saturated fat—coconut oil is approximately 90 percent saturated fat, which is more than double the amount found in lard. But the kind of saturated fat in coconut oil is different from that found in animal meat, eggs, dairy, and processed foods.

Coconut's saturated fat is a kind called medium-chain triglycerides (MCTs). MCTs are shorter in structure than the long-chain triglycerides (LCTs) found in animal products and processed foods. Since MCTs are shorter, they're more readily absorbed, turning into fuel for the brain and body more quickly than LCTs.[10] For this reason, MCTs are less likely to be stored by the body as fat. Studies also show MCTs suppress appetite and lower unhealthy cholesterol.

In the brain, MCTs work wonders, breaking down into compounds called ketones that neurons use for fuel when glucose isn't readily available. MCTs also improve cerebral circulation and fight age-related inflammation in neurons.[11] MCTs are so brain-beneficial that researchers are exploring how they might be used to treat Alzheimer's and dementia.[12]

Coconut oil is easy to incorporate into your diet by swapping out other cooking oils for this neutral-tasting option. The oil doesn't degrade in high heat, making it ideal for stir-frying, roasting, and other types of cooking, and is solid at room temperature, so it can be used as a butter alternative in baked goods.

Step Three: Yes, You Should Eat Carbs, Including One Type Many People Mistakenly Avoid

The brain needs a continual supply of glucose to function optimally. The best source of glucose comes from carbohydrates, which the body breaks down more easily into simple sugars than it does from fat or protein.

This doesn't mean you should load up on bread, pasta, cookies, and potato chips. Refined or simple carbs are toxic to cognitive health. They often contain too much sugar, causing a glucose spike that interferes with neuron function, and can shrink key brain regions involved in cognition, which leads to memory problems, difficulty thinking, and the increased risk of cognitive decline.

The type of carbs best for the brain are complex carbs like whole grains, vegetables, nuts, beans, and fruit, which contain more nutrients like fiber, vitamins, minerals, and antioxidants than refined or simple carbs. While complex carbs still have sugar, it comes from a natural source and breaks down more slowly than the stuff found in refined carbs. Your body also digests complex carbs more slowly because they are made up of longer chains of molecules, helping you to feel full and providing a lasting source of energy.

When it comes to complex carbs, there's a lot of confusion surrounding whole grains. While many packaged foods like bread, cereal, crackers, and frozen meals claim to contain whole grains, the products often also include refined carbs, sugar, and chemical additives. Processed "whole grains" are often low in fiber and healthy fat.

Which whole grains should you eat? Examples of healthy whole grains include brown rice, wild rice, whole oats, quinoa, amaranth, farro, buckwheat, barley, and millet. All are low on the glycemic index, a scale that ranks foods from 0 to 100 based on how they much they raise blood sugar, with 0 having no effect and 100 representing pure glucose, meaning it produces an instant and sharp spike.

Some contemporary trends like the Paleo and ketogenic diets shun whole grains, but I don't recommend doing so if you want to optimize brain function. While fruits and vegetables contain the glucose our brain needs, whole grains are the most concentrated source and are digested more slowly, providing a consistent sugar supply.

Whole grains are also rich in fiber, vitamins B and E, and other nutrients critical to cognitive function. The nutrients in whole grains help produce neurotransmitters, including the "feel-good" chemical serotonin (one reason people can feel euphoric after eating complex carbs).

Research also shows people who eat the most whole grains have a lower risk of age-related cognitive decline.[13] On the other hand, those who don't consume many whole grains are more likely to develop advanced cognitive issues and disease.[14]

Supermarket Shopping List:
12 Whole Grains for Better Brain Health

- Whole oats
- Quinoa
- Brown rice
- Wild rice
- Millet
- Farro
- Amaranth
- Bulgar
- Buckwheat
- Rye
- Spelt
- Barley

Step Four: This One Food Group Should Make Up Most of Your Diet

If you want to be healthy, smart, and slim, you'll follow food journalist Michael Pollan's golden rule and eat mostly plants. People who primarily consume plant-based foods have a lower risk of age-related cognitive decline, mental health problems, and neurodegenerative disorders, in addition to heart disease, obesity, stroke, cancer, diabetes, arthritis, and nearly every other chronic ailment. Countries where people's dietary habits are centered around plants also have some of the world's thinnest people.

Consuming a plant-based diet means eating foods grown out of the ground like leafy greens, fruits and vegetables, legumes, nuts and seeds, and whole grains. Calorie for calorie, plants, especially dark, leafy greens, are richer in vitamins, minerals, antioxidants, phytonutrients, and other compounds than any other food. Our brain needs these micronutrients to function optimally, even though most Americans don't consume enough of them.

In fact, only one in every ten people in the United States consumes enough fruits and vegetables to support their health, according to the CDC. Here are six plant-based foods to eat for a smarter, healthier brain:

1. **DARK GREEN VEGETABLES:** Kale, spinach, broccoli, swiss chard, collards, arugula, cabbage, watercress, mustard greens, bok choy, romaine lettuce, mesclun, endive, escarole, mixed greens, broccoli rabe.

   WHY YOU NEED THEM: If you eat only one vegetable today, make it a dark green one. Dark green veggies contain more vitamins, minerals, antioxidants, and phytonutrients per calorie than any other plant. They're rich in magnesium, a mineral critical to cognitive function, along with vitamins K, C, and E, lutein, folate, and beta-carotene—all needed to boost mood and mental alertness and prevent cognitive decline. Dark green veggies are also one of the few sources of glucosinolates, healthy compounds that fight oxidative stress in the brain, and chlorophyll, a green plant pigment that helps oxygenate and purify blood. Studies show consuming multiple servings of dark green veggies daily can fight neurodegenerative aging and decline, and increase brain function and performance.

2. **OTHER VEGETABLES:** Cauliflower, mushrooms, artichokes, brussels sprouts, green peppers, asparagus, avocado, bean sprouts, eggplant, cucumber, leeks, onions, zucchini, alfalfa sprouts, garlic.

   WHY YOU NEED THEM: Just because a vegetable isn't dark green doesn't mean it's not packed with brain-boosting benefits. Cauliflower, turnips, and brussels sprouts, for example, contain the same compounds found in broccoli and kale that fight oxidative stress.[15] Asparagus and brussels sprouts are also rich in folate, which our brain needs for neuron function, stress reduction, mood regulation, and disease prevention. In fact, every veggie listed above provides neuroprotective benefits, with the ability to fight disease and lift mood.

3. **ORANGE, YELLOW, AND RED VEGETABLES:** Acorn squash, carrots, red peppers, sweet potatoes, orange peppers, radishes, red cabbage, yellow peppers, butternut squash, pumpkin beets.

   WHY YOU NEED THEM: If you want to boost brain health, eat at least one serving of an orange, red, or yellow vegetable every day. These brightly hued veggies are concentrated sources of vitamins A, B, and C, along with beta-carotene and potassium, all which are vital to cognitive function, stress reduction, antiaging on a neuronal level, and the reduced risk of neurodegenerative disease and decline.

Colorful starchy vegetables like sweet potatoes and carrots also provide a healthy source of sugar for the brain. Sweet potatoes in particular contain vitamin E, which can trigger neurogenesis and help prevent cognitive diseases like Alzheimer's and Parkinson's. Like whole grains, sweet potatoes also help trigger mood-boosting serotonin.

4. **FRUIT:** Apples, blueberries, strawberries, raspberries, pears, oranges, grapefruits, melon, blackberries, pomegranate, lemons, grapes, watermelon, apricots, peaches, plums, pineapple, bananas, nectarines, cherries, cranberries, kiwis, mandarin.

   WHY YOU NEED THEM: While some fad diets eschew fruit as too high in sugar, many studies show that people who consume more fruit have healthier bodies and brains. Fruit does contain sugar, but in the natural form of fructose, not as an added or artificial sweetener. Fruit is also high in antioxidants that reduce oxidative stress and cognitive inflammation. Berries in particular contain antioxidants and plant pigments called flavonoids that improve memory. And blueberries may be one of the best brain foods, thanks to their ability to spur neurogenesis. Citrus fruits are also high in vitamin C and other micronutrients needed to prevent age-related decline.[16]

5. **LEGUMES:** Black beans, kidney beans, chickpeas, lentils, soybeans, edamame, lima beans, navy beans, black-eyed peas, cannellini beans, mung beans.

   WHY YOU NEED THEM: Beans and peas are the unsung heroes in the fight for mental fitness. Legumes are high in protein and fiber without the toxins often found in meat and dairy products. Legumes are also a rich source of folate and B vitamins, which are essential for optimal brain function. B vitamins are also critical to maintaining healthy levels of the mood-boosting neurotransmitter serotonin. Since B vitamins are water-soluble, our bodies can't store them, so we need to get them through food on a daily basis. Soybeans, edamame, and other legumes also contain antioxidants known as polyphenols that can help prevent dementia.[17] Like whole grains and vegetables, legumes are also a low-glycemic complex carbohydrate that provide a steady, healthy source of sugar to the brain.

6. NUTS AND SEEDS: Walnuts, almonds, cashews, Brazil nuts, sunflower seeds, hemp seeds, pistachios, pecans, pumpkin seeds, macadamia nuts, hazelnuts, chia seeds, pine nuts, peanuts.

WHY YOU NEED THEM: A recent study from China made major headlines when researchers proclaimed eating just two teaspoons of nuts daily could boost cognitive function in older people by as much as 60 percent.[18] That's because nuts and seeds, particularly sunflower seeds, almonds, and hazelnuts, are high in vitamin E, which helps fight oxidative stress, protect neurons, and reduce the risk of Alzheimer's.

Nuts and seeds contain other micronutrients and healthy fats shown to lower inflammation, reduce LDL cholesterol, and improve cerebral circulation. Studies even show consuming nuts on a regular basis can strengthen the brain wave frequencies associated with cognition, learning, memory, and healing.[19] Some nuts and seeds like walnuts, chia, hemp, and flax seeds are also rich in ALA. Walnuts are also a brain superfood, with research showing this nut can improve neuron messaging, preserve and boost memory, reduce inflammation, and trigger neurogenesis.[20]

Make Eating Plants Beneficial to Your Body

BUY ORGANIC: Whenever possible, eat organic, especially when it comes to fruits and vegetables. Conventionally raised produce contains harmful pesticides that can act like toxins in the brain, leading to brain fog, memory loss, weight gain, high blood sugar, and high cholesterol. The organic label also guarantees you won't be exposed to genetically modified ingredients, which could have neurotoxic effects.

GO RAW: High heat can cause enzymes and nutrients like vitamins B and C to degrade. When possible, eat raw veggies and fruits to maximize their living enzymes and micronutrients.

Step Five: Prioritize Plants for Protein—And Then Be Picky About Your Animal Products

Over the last decade, the number of studies conducted on plant-based diets has increased tremendously, along with evidence showing the health benefits.[21] This research has been largely comparative, contrasting the effects with omnivorous eating. The results have not proven beneficial for meat lovers, with studies finding numerous detriments to consuming animal products, including an increased risk of chronic illness, weight gain, low energy, brain fog, mood disorders, and much more. One study even found if we replaced just 3 percent of the animal protein we eat with protein from plant sources, we'd significantly improve mortality rates.[22]

What's so dangerous about meat, eggs, and dairy? Researchers point to harmful components like nitrosamines, which are carcinogenic chemical compounds found in meat, cheese, fried foods, and tobacco products. Studies show nitrosamines, also abundant in processed foods, increase the risk of neurodegenerative decline, particularly Alzheimer's.[23]

Animal meat is also rich in heme iron. While this nutrient is beneficial if you're anemic, consuming too much can build up in the brain and cause oxidative stress. People with Alzheimer's, Parkinson's, and other neurodegenerative diseases often have higher levels of heme iron.[24] Non-heme iron, found in whole grains, vegetables, beans, nuts, and fruits, has the opposite effect, reducing oxidative stress.[25]

There is also a growing body of evidence to show consuming animal protein increases systemic inflammation in the brain. One reason may be that meat contains pro-inflammatory chemicals that, when cooked, can multiply.

Animal products are also high in saturated fat, in the form of long-chain triglycerides (LCTs). While saturated fat is enjoying a resurgence thanks to ketogenic and Paleo diets, research shows LCTs are detrimental to brain health and can lead to memory loss, poor brain performance, and other cognitive problems.[26]

Animal dairy contains both saturated fat and sugar—a double blow to the brain. While the sugar found in dairy, lactose, is naturally occurring, we rarely consume cheese, yogurt, milk, butter, or cream on its own. Instead, we pair these products with refined grains (milk with cereal, cheese with crackers, butter with bread), more sugar (sugar-sweetened yogurt, ice cream, milk-based coffee drinks), or in the form of a processed food (cheese in pizza and pasta, butter in cookies and cakes).

Our body also can't easily digest lactose, which can increase inflammation and digestion issues that interfere with your gut-brain connection (more on this in Step Six). Casein, a kind of milk protein, can also trigger the body to produce too much mucus—one reason why the warm cup of hot cocoa I loved when I was sick as a kid never actually made me feel better. The pasteurization process used to kill bacteria in milk also lowers the number of vitamins, minerals, and other healthy compounds.

Modern-day animal meat and dairy also contain toxins, like antibiotics, hormones, steroids, and pesticides, that our ancient ancestors never consumed. Similar to chemicals in processed foods, toxins found in animal products can cause significant harm, accelerating cell aging, impairing cognitive function, and disrupting healthy gut bacteria.

You don't have to give up all animal products, but I do recommend limiting your intake significantly, with no more than one serving of meat or dairy per day, and buying organic whenever you can. The "organic" label ensures animals are fed nothing but organic, non-GMO grass or grains, and reduces your exposure to antibiotics and hormones.

Instead of consuming cow's milk, I suggest experimenting with any of the popular plant-based alternatives like coconut, almond, rice, soy, cashew, and oat milk. All are more easily digestible than cow's milk, without the same potential health hurdles.

If you don't want to give up dairy, try goat milk and yogurt, which are more popular than cow-dairy products outside the United States.[27] Research shows goat dairy contains more healthy fats than cow's milk, and it can also be gentler on the digestive system if you're lactose sensitive. I also like plain, unsweetened, organic Greek yogurt because it's high in protein and free of sugar and additives.

The Case for Plant Protein

Plant protein isn't healthy only because animal products are so harmful by comparison. Consuming more plant protein can help protect your brain and reduce blood sugar, inflammation, bad cholesterol, and blood pressure.[28] Studies show populations around the world who consume the most plant protein live longer, enjoy a higher quality of life, preserve their mobility, are generally happier, and retain more of their cognitive power into their advanced years. Even top athletes like Tom Brady and Serena Williams can meet their protein needs eating only plants. Here are five excellent sources of plant protein, with what you need to know about each:

1. **SOY**, including tofu, tempeh, edamame, soy milk, miso, whole soybeans, soy nuts.

   Soy is a complete protein, meaning it contains all nine amino acids. Our body is also highly adept at absorbing the protein from soy. One-half cup of tempeh contains approximately 14 grams of protein, while the same amount of tofu has 10 grams, edamame 9 grams, and soymilk 4 grams (the same amount found in cow's milk).

   Soy also is rich in many brain-healthy nutrients, including B vitamins, zinc, calcium, Coenzyme Q10, potassium, and magnesium. Soy also contains isoflavones, plant compounds that studies suggest improve cognitive function. While isoflavones are naturally estrogenic, they don't actually act like estrogen in the body, meaning they won't raise breast cancer risk or feminize men who consume high quantities of the plant protein.[29] Fermented options like tempeh and miso also have healthy bacteria that benefit the gut and the brain.

   While there's nothing inherently unhealthy about soy, you still need to be careful when you eat it. Ninety-five percent of conventionally grown soy is genetically modified, which could cause neurotoxic effects—avoid the risk by always buying organic. Also steer clear of the dozens of highly processed soy products on store shelves, including soybean oil, soy burgers, soy margarine, soy cheese, and soy isolate, found in protein powders and shakes.

2. **LEGUMES**, including lentils, chickpeas, black beans, kidney beans, lima beans, fava beans, pinto beans, cannellini beans, green peas.

   You'll get at least 8 grams of protein in every half cup of lentils and beans, while green peas (like the kind you'd find in peas and carrots) boast around 4 grams of protein per half cup. Legumes aren't a complete protein, meaning they don't contain all nine amino acids, but if you eat a balanced diet with whole grains and other plants, you'll have nothing to worry about. Beans also contain fiber, which helps to nourish gut bacteria, regulate cerebral circulation, and trigger weight loss.

3. **WHOLE GRAINS**, including quinoa, buckwheat, oats, amaranth, millet, brown rice, wild rice, spelt, rye, barley.

   Consider quinoa and amaranth the power proteins at the top of the plant pyramid. Both are complete proteins, with 4 or 5 grams of protein in every half cup, respectively.[30] Whole oats and brown rice aren't complete proteins, but they still contain about 5 grams of protein in every half cup. Similar to soy and legumes, whole grains are also high in fiber.

4. **NUTS AND SEEDS**, including almonds, walnuts, cashews, chia seeds, hemp seeds, pecans, pistachios, peanuts, sunflower seeds, pumpkin seeds.

   The ordinary peanut is a protein powerhouse, with 7 grams in every ounce, which is the equivalent of about twenty-eight nuts. (While peanuts are technically a legume, many experts group them with nuts, culinarily and nutritionally.) Next up for protein among all nuts are almonds, which have 6 grams per ounce, the equivalent of about twenty-three nuts. Most others like cashews and pistachios contain around 4 to 5 grams per ounce, with pecans and macadamias trailing the pack, at 2 and 3 grams, respectively.

Seeds top nuts in nutrient density, including protein. Hemp seeds, for example, pack a whopping 10 grams of protein in every ounce. The same amount of pumpkin or flaxseeds boast about 5 grams per ounce, with 4 grams for chia seeds.

Both seeds and nuts are rich in a wide range of nutrients, including fiber, vitamin E, and healthy ALA fat.

5. **VEGETABLES**, including potatoes, broccoli, mushrooms, spinach.

   Few people think about vegetables when it comes to maximizing muscle power with diet. But plenty of plants pack protein in quantities that can help you meet your daily needs. For example, a large baked sweet potato provides 4 grams of protein while one-half cup portobello mushrooms contains 4 grams. You'll find 3 grams in one cup of broccoli and the same amount in one-half cup spinach.

Step Six: Eat to Feed Your Gut-Brain Axis

The "gut-brain axis" is a relatively simple term for a complex process in the body that significantly affects brain health and function. Only recently discovered by doctors, the gut-brain axis describes the line of communication that exists between our brain and gastrointestinal tract. Basically, what happens in our gut affects the brain's physical, mental, and emotional functions, influencing neurotransmitter production, behavior, pain regulation, and stress modulation.[31]

The gut-brain axis is rooted in the body's microbiome, a community of 100 trillion microorganisms like bacteria, fungi, protozoa, and viruses that live on and inside you. This community is so big, it contains over one hundred times more genes than the human genome and weighs up to five pounds—twice that of the brain. In recent years, doctors have discovered the microbiome helps control a number of physical and cognitive functions and plays a role in the development of heart disease, diabetes, cancer, and neurodegenerative disorders like Alzheimer's and Parkinson's.

Our microbiome is made up of both good and bad bacteria—and it's a very delicate balance between the two. Develop too much of the bad variety and you can throw your entire body out of balance, increasing the risk of weight gain, depression, anxiety, high cholesterol, high blood sugar, fatigue, intestinal distress, and other ailments.

The best way to ensure a healthy microbiome balance is by prioritizing plant-based foods. Research shows adopting a diet of just plants for only five days can start to diversify the bacteria in your gut and even trigger genetic changes within the microbiome.[32] It also helps to consume a wide variety of foods, especially those that contain fiber, which helps feed healthy bacteria.

Avoiding processed foods and choosing only organic will also help nourish your microbiome. That's because common additives like sugar, hydrogenated fat, emulsifiers, artificial sweeteners, and food dyes can kill healthy bacteria while feeding the bad kind, allowing them to multiply. Pesticides, antibiotics, hormones, and steroids used in conventional animal ranching are also deadly to healthy bacteria.

Some foods also contain healthy bacteria, known as probiotics. These foods include fermented items like tofu, tempeh, raw sauerkraut, kimchi, kombucha (a fermented tea drink), kefir (a fermented milk drink), and plain unsweetened organic yogurt. You can also take a probiotic supplement, which you'll learn more about in chapter five.

Step Seven: Try Intermittent Fasting

If you've read anything about weight loss in the last few years, you've likely stumbled on the idea of intermittent fasting. Intermittent fasting doesn't mean you don't eat for days on end, but that you fast twelve to eighteen hours between meals, usually between dinner and breakfast or lunchtime the next day.

Intermittent fasting isn't a newfangled fad, but a well-researched lifestyle modification that can help you live a longer, healthier, smarter life. Research shows people who practice regular intermittent fasting have lower body fat, resting-heart rates, blood-sugar and insulin levels, unhealthy cholesterol, and bad blood fats.[33]

Intermittent fasting has such profound effects because it causes the body to shift metabolic pathways, from relying on blood sugar for fuel to tapping into stored fat for energy—this is why the habit can cause weight loss. When you don't eat for hours, your cells think they're starving and go into survival mode, expunging unhealthy mitochondria and replacing them with new ones. The body also doesn't produce insulin while fasting and instead increases levels of human growth hormone, which stimulates cell growth and regeneration along with the release of the neurotransmitter norepinephrine, which helps fight depression and other mood disorders.

For the brain, intermittent fasting has been shown to improve memory, focus, learning, and overall executive function.[34] The habit also can slash oxidative stress and cognitive inflammation, trigger neurogenesis, and increase neuroplasticity, or the ability of your brain to change.

Intermittent fasting can also help you become more mindful around food. Many of us eat out of habit or because we're bored, upset, or stressed. Fasting, though, forces you to pay attention to what and when you eat and make more conscientious choices.

5 Steps to Starting Intermittent Fasting

1. **START WITH 12 HOURS.** Don't try to tackle a sixteen-hour fast on Day One. Make it your goal to fast for twelve hours by finishing dinner by 7 or 8 P.M.—no snacking afterward—and not eating again until after the same time the following morning. Once you get comfortable with a twelve-hour fast, increase your fasting duration by thirty minutes to an hour until you feel comfortable lasting sixteen hours between meals.

2. **FASTING DOESN'T MEAN DEHYDRATION.** Consuming lots of water, decaffeinated coffee, and unsweetened tea can help keep you feeling full while revving your metabolism, blood flow, and energy. Just be sure to avoid beverages with calories or sugar, which will kick you out of a fasting state.

3. **FOCUS ON QUALITY FOODS.** If your last meal of the day is high in refined carbs and sugar with little fat, protein, or other nourishing nutrients, it'll drive up your appetite and you'll have a more difficult time sustaining your fast. For this reason, make sure to eat a healthy dinner high in fiber, protein, and healthy fat so that you're able to maintain your fast through the following morning.

4. **REMEMBER THAT YOU WILL EAT AGAIN.** Hunger is a normal sensation when you start fasting, but don't give up—remember that you will eat again. After a few days fasting, your body will also start to adapt to your new routine and you'll feel less hungry over time. Finally, there's nothing wrong or harmful about hunger: It means your body is working to clear unhealthy cells while your digestive system is getting a chance to heal.

5. **CONSULT A PHYSICIAN BEFORE FASTING.** Some people like pregnant women and those with type-1 diabetes, cancer, or eating disorders shouldn't be intermittent fasting. Be sure to check with your doctor before starting a fasting regimen.

The Wonders of the Mono Meal

Digesting food, no matter how healthy it is, is work for your digestive tract. It can take a full hour to digest complex carbs like fruit and vegetables while high-protein foods like fish, soy, and beans can take up to three hours to process. If you eat a wide variety of foods in large quantities, it adds more stress to your digestive tract.

This is why I love the concept of the mono meal at night. A mono meal is comprised of a single food item only—think a plate of steamed or raw vegetables, a bowl of fresh fruit, some plain oatmeal, or roasted sweet potatoes. You shouldn't eat a mono meal every night, but incorporating one once or twice per week will give your digestive tract a healthy break and retrain your taste buds to enjoy food with less sugar and additives.

Putting It All Together: The Best Way to Follow a Better Brain Diet

No matter what your nutritional plan is now, you can enhance your diet to improve your brain power and performance. I've worked with people who, when we first met, had unbelievably poor diets and swore there was no way they could start eating mostly plants and other whole foods. They were amazed, then, when they accomplished the feat. You can do this too, not by changing everything at once but by learning strategies to improve your nutrition in small, incremental ways. Here are four to keep in mind as you start to modify your diet.

1. **EAT LIKE YOU'RE IN THE MEDITERRANEAN.** The Better Brain Diet is similar to the popular Mediterranean diet in the sense that both are rich in whole grains, vegetables, fruits, legumes, healthy oils, and nuts and seeds, with small amounts of seafood and poultry. The Mediterranean diet is a great starting point, but what's even better for cognitive health is the MIND diet, short for Mediterranean-DASH (Dietary Approaches to Stop Hypertension) Intervention for Neurodegenerative Delay.

   The MIND diet became popular in 2015 after researchers at Rush University Medical Center discovered people who adhered to the plan reduced their risk of Alzheimer's by 53 percent. Even those who only casually followed the diet cut their chances of developing the disease by 35 percent.[35]

   What I like best about the MIND diet is that it's prescriptive in how many servings you eat, making it easy to follow. On the MIND diet, you eat a daily minimum of two to three servings of vegetables, including one dark, leafy green, and three servings of whole grains. On a weekly basis, you eat at least two servings of berries, four servings of legumes, five one-ounce servings of nuts and seeds, one serving of seafood, and two weekly servings of poultry. All serving sizes are standard, based on USDA guidelines.

   The MIND diet also replaces butter, margarine, and other forms of added dietary fat with olive oil, limits the consumption of red meat to fewer than four servings per week, and includes little to no fried foods, cheese, refined grains, or items with added sugar.

2. **MODIFY TO CHANGE YOUR BRAIN AND LOSE WEIGHT.** I love the MIND diet as a great place to start if you're concerned only with

your cognitive health, but the plan isn't designed for weight loss. After working with football players and hundreds of others who wanted to lose weight and boost their brain, I decided to modify the MIND diet to help people accomplish both.

The Better Brain Diet speeds weight loss by advocating a greater daily consumption of vegetables, fruit, legumes, and nuts and seeds. I also encourage swapping out red meat and poultry for plant protein, which is more conducive to sustainable and lasting weight loss. The diet also includes coconut oil, which suppresses appetite and stimulates fat loss.

On the Better Brain Diet, you eat three servings daily of green vegetables, including one orange, yellow, or red veggie. On a daily basis, you also eat two servings of seasonal fruit (berries are optimal for brain health), one serving of legumes, two servings of plant protein or seafood, one serving of whole grains, one serving of nuts or seeds, and three servings of healthy oils like coconut, olive, flaxseed, or hemp oil.

While the MIND diet allows for a daily glass of wine, I don't include any alcohol on the Better Brain Diet for two reasons. First, alcohol is empty calories and can lead to overeating. Second, having seen so many brain scans of people who've consumed alcohol, I simply can't recommend it as a daily part of any brain-healthy plan. However, if you've never had a concussion or other brain injury and don't suffer from psychiatric or neurological issues, feel free to have a weekly (not daily) glass or two of wine, which contains some healthy nutrients, including the antioxidant resveratrol.

3. **EXPERIMENT, TRY, AND MODIFY.** The Better Brain Diet works only if you work to make it your own. You don't have to like every vegetable or food listed in this chapter, but you do need to keep an open mind, experiment, and be willing to discover new foods. Remember that items within the same food group can taste distinctly different, and you shouldn't

rule out one category until you've tried various types and preparation methods.

You can also experiment and modify the number of suggested daily servings to some degree. For example, if you don't want to eat fruit every day, consider increasing your intake of colorful veggies. Or if you can't stand beans, substitute more nuts and seeds to get a similar amount of protein and fiber.

4. **START BY WRITING IT ALL DOWN.** Before you start any new nutritional approach, it helps to know which areas of your current diet are working and which may need a slight overhaul. Before you begin the Better Brain Diet, take a few days to journal what, when, and how much you eat. Be honest—no one needs to see your food diary other than you.

   After a few days to one week, review your diet to find out which food groups you consume frequently, which ones you overlook, and the times of day you're most likely to overeat, crave sugar, or make other unhealthy choices. This can help identify good habits you already have and patterns that may need to change.

   If your diet is not great now, don't feel overwhelmed or discouraged. Think instead about how much power you have to improve your cognitive potential and overall health by making small improvements to your nutritional choices.

Paul's Story

LOSING 100 POUNDS ON THE BETTER BRAIN DIET AND GAINING COGNITIVE CONTROL

Paul, a fifty-six-year-old accountant from Southern California, came to work with me several years ago because he felt his stress and anxiety levels were out of control, and he had gained a lot of weight as a result. Paul was at least one hundred pounds overweight when we first met, and as soon as he started telling me his history, I began to see why.

Paul is married with four children, which, in itself, means he faces the common stressors many men do to be a good father and husband, and the financial provider for his family. In addition, Paul also has a three-hour commute each way to and from his office in Hollywood on one of the busiest freeways in California, the 405.

Once Paul got to work, he started the day with pastries, muffins, or some other refined carb offered by his office. He told me he would often just go into autopilot eating mode, consuming anything in his path that might give him some sense of comfort, including the free ice cream in his office cafeteria or potato chips from the vending machine. For lunch, he frequented all-you-can-eat places or had working lunches with clients or colleagues when he didn't pay much attention to what he was consuming. After a long day at work and another horrific commute home, Paul would stop for a snack at McDonald's or Taco Bell before eating dinner with his family. Finally, he treated himself to a nightly glass of wine or a martini before heading off to bed.

When I first started working with Paul, I asked him the same question I like to ask all my clients: what thoughts or perceptions do you have about the brain? Paul's response was typical. He told me he thought his brain was just like any other organ and that we all pretty much had the same brain, just like we have similar livers, kidneys, and gallbladders. In his mind, his brain was no different from mine, his wife's, or any of his colleagues in the office. This belief led him to assume we all have the same built-in willpower when it comes to food and food cravings—those who can't control their behavior around food are just weak, because they've lost their mental willpower.

While Paul is right that we have similar brains physically, each of us is cognitively unique, with individual genetics, life experiences, varying degrees of toxin exposure, and different histories of falls, hard hits to the head, and other mild brain injuries that can influence our ability to function optimally, mentally and emotionally.

In other words, what Paul perceived as his own weakness—an inability to exert willpower around food—had less to do with mental weakness and more about parts of his brain that might have not been functioning optimally. Making matters worse was that he was consistently eating processed foods high in sugar and toxins and gaining weight as a result, both of which were further supporting a vicious cycle of reduced willpower and poor nutritional choices. But Paul's brain could change, and I knew it was time for us to flip the switch.

The first thing I encouraged Paul to do was write down everything he ate. If he grabbed a single potato chip out of a bag, I wanted to know about it. This helped him to see exactly which types of foods he was eating, how much, and when—and how often he just mindlessly ate whatever stood in his path or came into sight. For example, he would often snack while cooking before realizing, through his journaling, just how much he was eating before he even sat down to the dinner he was making.

Next, I recommended that Paul stop eating all processed or manufactured foods—Step One in the Better Brain Diet. He cleared his house of his two trigger foods—chips and bread—and asked his family not to bring either home until he got his stress and cognitive willpower under control.

In cutting out processed foods, Paul also eliminated nearly all the added sugars he had been consuming. He also stopped drinking diet soda, which cut out the artificial sweeteners known to be toxic to our brain and waistline. In the place of diet soda, he began drinking more water, which helped him boost his metabolism and kept him feeling full throughout the day.

Slowly, Paul started introducing fresh fruits and vegetables into his diet, eventually making organic produce his primary source of calories for the day. At the same time, he cut out animal fats like butter and began treating meat more like a condiment than a main dish. Today, Paul still loves his favorite food, filet mignon, but he now enjoys a small portion on special occasions rather than indulging in a twelve-ounce steak on a regular basis.

Eventually, Paul also bought a food scale that allowed him to calculate the weight, calories, nutrients, fats, carbs, and proteins of everything he ate. While you don't have to do this, the scale initially helped Paul figure out exactly how much he was eating. Instead of a restrictive burden, Paul's food scale became a tremendous tool, arming him with the knowledge to make smarter food choices.

Paul loves to cook, which helped him to discover new ways to make food taste more flavorful without sugar, animal fats, unhealthy oils, heavy cream, or other additives. Instead, he began experimenting with herbs and seasonings, adding garlic and turmeric to savory dishes and cinnamon and nutmeg to steel-cut oats, roasted sweet potatoes, and mashed yams.

With the help of his wife, Paul also started to prioritize eating healthy brain foods like walnuts, avocados, blueberries, strawberries, and green veggies on a daily basis. I outlined a supplement program he could follow that would help lower his stress while boosting his brain health and weight-loss efforts, and he began a walking routine that eventually led him to running, swimming, and combat-style group exercise classes.

Today, Paul is one hundred pounds leaner than when we first started working together—and he is happier and healthier. Through diet and other lifestyle choices, he has effectively doused his stress and anxiety and overhauled his ability to focus, think clearly, and stay sharp. A lean man with no health issues, Paul doesn't necessarily need to follow the Better Brain Diet so rigorously anymore, but he chooses to do so because he truly enjoys what he eats—and has never felt better as a result.

KRISTEN'S TIP: Stress eating is a real thing, which you can see from Paul's story. Become more mindful of what you consume by eating set meals rather than snacking. Don't eat at your desk or in front of a computer or TV. Instead, take the time to enjoy and savor your food.

4

THE BETTER BRAIN WORKOUT

I've always loved to exercise, from gymnastics camp to tennis lessons to the day when my parents surprised me with a dapple-gray pony named Razzmatazz after I showed a sincere interest in riding at age seven. For the next ten years, I immersed myself in equestrian culture, spending every day at the barn, riding, jumping, and training to compete at horse shows throughout the Midwest. I would ride until my muscles were sore, my lungs were taxed, and my horse was dripping with sweat—and then I'd ride a little more. And I loved every minute of it.

As my riding skills advanced, I started show jumping competitively. At that time, I found nothing more exhilarating than jumping my thoroughbred racehorse, Lexington, over a five-foot fence at top speed and racing toward the finish line. The adrenaline got me hooked on exercise, fueling a forty-year addiction to move my body on a daily basis.

After I stopped showing horses, I had to channel my competitive nature into other forms of exercise. I started running, going to the gym, swimming, boxing, cycling, jumping rope, and trying other sports to see which could simulate that endorphin high. I've tried everything from golf, basketball, and rowing to Pilates, plyometrics, and surfing. To this day, I do some sort of daily physical activity, preferably outdoors, weather permitting. But I'm not dogmatic about it, and if I miss a day, I don't beat myself up over that.

Outside of competitive sports, running is my favorite activity—for me, it doubles as a form of physical activity and moving meditation. I tune in to the sounds of the birds, people along the boardwalk, or cars passing, and I plan my day. Running helps organize my mind. It's also very easy to lace up my shoes and cruise out the door. Then I'm home in an hour and ready to tackle my day.

If you aren't active now or work out only occasionally, the idea of exercising seven days a week may seem overwhelming or even obsessive—and what I do might sound totally over the top for some. But it really isn't. I'm lucky in that I gravitate toward activities that feel more like competitive games and challenges than actual exercise, so the negative associations that some have with exercise aren't there for me. These activities exist for everyone, and it's never too late to discover something new you love to do! I promise, if you start exercising and take the time to find the ways your body likes to move like I have, working out will become something you want to do, not something you have to do.

The reason I want you to embrace exercise is that it's one of the most powerful ways to biohack your brain. It will make you smarter and sharper, and it will protect your cognitive function as you age.

Exercise Maximizes Blood Flow to the Brain

One of the things that surprised me the most after I started working at the Amen Clinics and seeing hundreds of brain scans was the difference in cerebral circulation between those who worked out and those who didn't. People who exercised had dramatically more blood flow to the brain and less damage as a result than those who didn't. Healthy blood flow was helping their brains perform faster and more efficiently and staving off cognitive decline.

For many, it may be difficult to comprehend why cerebral circulation is such a big deal to cognitive function and why exercise is the golden key to boosting brain blood flow. Consider this: The human brain contains around four hundred miles of blood vessels, laced inside a space measuring only 1,200 cubic centimeters. To get blood deep inside the brain's vascular network, the heart has to be strong, and your arteries and veins need to be open for blood to flow. The best way to increase your cardiovascular health is to exercise, which trains your heart and turns your blood vessels into superhighways—smooth, wide, and fast. Older people who exercise regularly have vessels that appear as young and as healthy as people half their age, according to research.

You don't need to do a marathon to magnify cerebral circulation. A study found that older women who walked thirty to fifty minutes several times weekly improved brain blood flow by as much as 15 percent in three months.[1] On the other hand, not exercising in some capacity for even just ten days can reduce cerebral circulation by as much as 30 percent.[2]

The more blood you can pump upstairs, the more oxygen, sugar, and other nutrients your brain receives, helping you to react, process, think, remember, learn, and focus better. Magnifying cerebral circulation also boosts brain volume, strengthens synaptic connections, helps manufacture vital proteins and hormones, clear toxins that can lead to dementia, and grows new brain cells.

BEST EXERCISES FOR CEREBRAL CIRCULATION: Studies show sustained aerobic exercise that elevates your heart rate for a period of time—think running, cycling, and swimming—are the most beneficial. But resistance training has merits, boosting circulation to the limbs while building more muscle mass. The more muscle you have, the more places your body can pump blood, consequently reducing pressure on arterial walls. Yoga has been shown to lower blood pressure and bump up cerebral circulation.[3] Walking also increases brain blood flow, especially if you walk fast enough to raise your heart rate. There's another advantage to a brisk walk: when your feet hit the ground, the impact triggers pressure waves to oscillate through arteries, further increasing cerebral circulation.[4]

Redefining What It Means to Work Out

Whenever you raise your heart rate or challenge your limbs or lungs, you're getting exercise, whether you're gardening, hiking, or even doing household chores.

While some people take for granted the physical activity their body is capable of doing, not everyone has the same options for exercise due to injury, age, chronic pain, or degenerative conditions. However, I truly believe that there *is* something for everyone. If you have trouble standing or walking, for example, or if you're confined in a cast or brace, consider chair aerobics, which you do sitting while moving your arms and/or legs. You can also do many yoga poses in bed, use hand weights at home, or strengthen muscles sitting with resistance bands. Head online to find instructional videos on how to do these activities. Just be sure to check with your doctor or physical therapist before adding any new physical activity.

The Fastest and Most Effective Way to Get New Brain Cells

If there's one cognitive concept that intrigues all my clients, it's neurogenesis, the ability to grow new brain cells. Who doesn't want more brain cells? I'll tell you what I tell them: if you want to grow new neurons and increase your cognitive aptitude and intellect, you need to do some sort of aerobic exercise, which research shows is the most effective way to stimulate neurogenesis.

Scientist Fred Gage of the Salk Institute for Biological Studies and colleagues at the Sahlgrenska University Hospital in Sweden led pioneering research that helped discover that adults can grow new brain cells in the hippocampus, the area of the brain important in learning and memory.[5] Dr. Gage and his colleagues found that mice that had access to a running wheel stimulated hippocampal neurogenesis, neuroplasticity, and new learning more than mice that didn't have access to a running wheel.[6] This means the birth of neurogenesis, from a research standpoint, was physical activity.

Since then, studies have shown that aerobic exercise can double or even triple the number of new neurons in the hippocampus, the part of your brain responsible for memory and learning.[7] While we don't fully understand how exercise has this impact, what we do know is physical activity stimulates the brain to produce brain-derived neurotrophic factor (BDNF), a protein shown to regulate neurogenesis. Exercise also causes the blood in the brain to release certain proteins that trigger neurons to form in the hippocampus.[8]

BEST EXERCISES FOR NEUROGENESIS: Unlike cerebral circulation, which you can increase simply by moving your body more, neurogenesis is stimulated by specific forms of exercise, primarily running and other sustained aerobic activity. In animal studies, the brains of rats who ran on a treadmill for six to eight weeks showed the biggest uptick in new neurons—more than their littermates who did sprints or other high-intensity intervals for the same amount of time. As for rats who only lifted weights (i.e., climbing up a vertical ladder with weights), their brains showed no discernible new neurons over a sedentary control group.[9]

Work Out for a Smarter, Bigger Brain

Study after study shows people who exercise on a regular basis outperform couch potatoes on cognitive tests.[10] But how does exercise make us smarter? Boosting cerebral circulation and stimulating neurogenesis helps, but physical activity has benefits beyond this.

Working out regularly increases the hippocampus, the part of the brain responsible for memory and learning. The bigger your hippocampus, the better your brain can retain memories and learn new information and skills. Building hippocampal volume also protects the brain from mood problems like depression and neurodegenerative disorders like Alzheimer's.

Similar to other areas in the brain, the hippocampus shrinks with age—one reason we struggle with memory problems and lower cognition as we grow older. But studies show exercise prevents and can even reverse age-related shrinkage. It's actually one of the few "proven" methods, according to researchers, to retain hippocampal size and function.[11]

Physical activity also increases gray matter across the brain. What's so great about gray matter? More of this type of tissue improves the brain's overall ability to think, reason, and remember. Having thicker and healthier gray matter can also prevent Alzheimer's and other neurodegenerative disorders.

Doing everyday activities like housework and gardening can build gray matter, too, according to studies, which show that people who are active outside a gym have more of this type of tissue than those who aren't.[12]

What about white matter? Turns out physical activity works wonders there, too, increasing the volume and connectivity of nerve fibers, which makes up more than half our overall brain.[13] Exercise also improves connectivity between the brain's left and right hemispheres, increasing creativity, language skills, memory retrieval, concentration, and muscle coordination.

BEST EXERCISES TO BOOST BRAIN SIZE AND GET SMARTER: You guessed it: sustained aerobic activities like running and walking are also best at building the hippocampus and other gray matter. While recent studies show yoga can increase hippocampal size,[14] the jury is still out on whether lifting weights and other forms of resistance training can actually trigger gray matter growth.[15]

When it comes to brain connectivity, distance running is the best way to increase the number and variation of the brain's synapses. Scans of runners' brains show greater connectivity between networks associated with executive function and motor control, according to multiple studies.[16] That's because when you jog or run, you force your brain to multitask as you navigate, respond to stimuli around you, analyze road conditions, and engage in sequential muscle-motor skills.

Fight Brain Stress, One Step at a Time

If you've ever ended a tense day with a long walk or sweat session in the gym, you know how a good workout can cut stress, calm nerves, and leave you feeling better about life. That's because exercise triggers a cascade of physical effects that impacts the sympathetic nervous system. While exercise produces the stress hormone cortisol, it energizes the brain rather than detracting from your cognitive function. Working out also increases production of endorphins and the neurotransmitters dopamine, serotonin, GABA, and noradrenaline, all of which lift mood and reduce stress.

Physical activity also conditions the body to better regulate cortisol over time.[17] Studies show animals with unlimited access to exercise have lower fight-or-flight responses than those who can't move as much as they want.[18]

Here's a frightening research fact about exercise and stress: studies show that if you don't work out, your neurons can even change shape, sprouting new branches that will make you more susceptible to anxiety and tension.[19]

BEST EXERCISES TO REDUCE STRESS: Any movement that makes you happy is a great way to manage stress. Forget what your friends like to do or a recommended workout you don't enjoy; forcing yourself to do something you don't like can increase stress, upending the benefits of that exercise.

Beyond finding something you like, research shows people who work out in groups, whether in a running club, yoga class, or a dance lesson, reduce stress more than those who do solitary activities. Scientists say the social benefits of group exercise, along with the emotional and mental support they can provide, bolster exercise's stress-reducing effects.[20]

Low-impact activities like yoga, tai chi, and Pilates that combine movement with breath work can bring deeper levels of calm and serenity. (For more on the benefits of deep breathing, see page 158.) Nontraditional forms of exercise like gardening have also been shown to be effective at fighting acute stress. In fact, one study found gardening cut stress more than reading quietly inside.[21]

The NFL Story

HOW THE RIGHT EXERCISE CAN OVERHAUL AN ATHLETE'S BRAIN

I'm always encouraged to see people step outside their comfort zone and try new things. That was what happened when I started working with Lance Zeno, a former offensive lineman for the Cleveland Browns and Green Bay Packers. As a collegiate and pro player, Lance had exercised his entire life, but his primary focus was lifting weights, with some light aerobic exercise.

After I saw his brain scan, I knew Lance needed to reduce the electrical firestorm inside his head. He was also having problems sleeping at night and felt stressed often. Like most players, he was also concerned about the collective toll playing high school, collegiate, and professional ball had taken on his cognitive function. When I met him, Lance was pursuing a graduate degree in education and was finding the work more challenging than he remembered his difficult undergraduate course load at UCLA ever was.

In addition to dietary changes and a new supplement routine, I suggested Lance start incorporating yoga, stretching, and meditation into his workout regimen. He started by taking two yoga classes a week, finding the practice a little out of his comfort zone at first. But once he began to feel less stressed and more focused, and he was able to sleep through the night, he was sold on the practice.

After tracking his progress for several months, Lance told me that yoga had helped to boost his physical, mental, and emotional health more than any other exercise he had done in the past. With yoga, he had more energy, better balance, and less joint pain, and he was sharper in his studies and everyday life. Adding yoga to his workout routine had also helped him lose thirty pounds.

Today, Lance still practices yoga two to three times per week, adding that he can't remember when he didn't leave a class feeling recharged, recentered, and more positive. He's even started teaching yoga to the at-risk teens he works with in a youth center outside of Los Angeles. As proof of yoga's power, he points to one teen, a former gang member, who compared the euphoria of yoga to street drugs, but with an effect that's gentler on overall mood and longer lasting.

KRISTEN'S TIP: Experiment with new types of exercise, even if it's outside your comfort zone or you assume you wouldn't like it. Sometimes the activities we are most apprehensive to try are the ones we need the most.

Move Your Body, Change Your Mind

Reducing stress isn't the only way exercise can overhaul mood. Physical activity can also relieve feelings of sadness, restlessness, boredom, dissatisfaction, low self-esteem, and even depression. In fact, a regular exercise routine may treat some cases of clinical depression just as effectively as prescription antidepressants.[22] Exercise has also been shown to treat ADHD, working similarly to prescription drugs like Ritalin and Adderall by stimulating neurotransmitters that increase concentration.

When Exercise Becomes Part of the Problem, Not the Solution

Exercise is one of the best ways to counter chronic stress, but working out too long, too frequently, and/or too intensely can have the opposite effect. If you're already hitting your red line, very long or intense workouts can send you into cortisol overload.

Symptoms of unhealthy cortisol levels include sleeplessness, fatigue, weight gain (no matter how much you exercise), anxiety, and difficulty concentrating. Talk with your doctor about getting your cortisol levels checked if this sounds like you. If you're exercising too much or too intensely, find alternatives outside of exercise to lower stress, like meditation and breath work (see chapter seven).

If you've ever finished a good aerobic workout, you know just how great physical activity can make you feel. This "runner's high" happens after exercise stimulates the brain to release endorphins, along with serotonin, dopamine, and norepinephrine, all of which can help us feel more satisfied, positive, and peaceful. Physical activity also triggers the increase in brain derived neurotrophic factor (BDNF), the same molecule responsible for stimulating neurogenesis that also helps us feel happier and more optimistic.[23] Even exercising for five minutes can help you feel better about yourself, according to studies.[24]

BEST EXERCISES TO BOOST MOOD: Whatever workout you like is the one that will improve your mood the most. And good news if you prefer weights: both aerobic exercise and resistance training have been shown to produce feel-good effects on the brain.

One word of warning: don't start out too quickly or intensely. Studies show that beginning a workout at a pace or an intensity that makes it too difficult to talk can postpone exercise's mood-boosting effects by about thirty minutes.[25]

Double the Benefits to Your Brain by Exercising Outside

While a stroll around your office or on a treadmill is fantastic, you'll reap the benefits of outside light and a boost in vitamin D production if you work out where you can see trees, fields, lakes, rivers, or other green or blue space. Known as "green exercise," working out in nature has been shown to decrease anger, anxiety, depression, sadness, and stress more so than exercise in urban or suburban environments.[26] Brain scans also show people who are active in nature have less cortisol and lower levels of activity in the brain where negative thoughts and ruminations occur.[27]

Frank's Story

HOW WALKING AND DANCING UPENDED HIS COGNITIVE PROBLEMS AND HELPED HIM LOSE 100 POUNDS

When Frank first came to see me, he was struggling with bipolar disorder, depression, and weight gain, among other issues. At the time, the forty-three-year-old didn't exercise—despite owning a treadmill, rowing machine, and NordicTrack, all of which he primarily used as racks to hang up his clothing instead of for exercise. Frank had even purchased a book with the playful title *Help Lord—The Devil Wants Me Fat* to see if it would inspire him to get active, but the tome made him laugh more than it motivated him to his treadmill. All the while, he rationalized his lack of extracurricular exercise with the fact he walked at his job as a zoning inspector and was getting his exercise there.

After hearing about his symptoms and understanding how his brain was wired, I recommended Frank adopt some kind of sustained aerobic exercise—not just sporadic walking at work—to balance his mood and counteract the mania he was feeling. I suggested Frank try walking outside, so to get him inspired we started conducting our sessions walking together outdoors.

From there, Frank took the initiative and began walking outside two to three times per week, first for thirty minutes at a time, which he quickly increased to forty-five minutes, then a full hour. He started wearing a fanny pack and bringing an ancient Discman on his walks, using music to motivate him to go farther. He'd play games with himself, seeing if he could get to the next intersection by the time a new song started. Eventually, Frank began walking for up to three hours at a time.

Frank also started taking Zumba classes, where he usually was the only man in the class. This delighted the teacher, who positioned him in the front of the class where everyone could see him, which motivated him to try harder. Realizing that he loved the physical freedom dance provided, he eventually bought a DVD set so he could Zumba at home.

Physical activity significantly decreased Frank's symptoms of depression and mania, quelled his negative thoughts, and improved his mood more effectively than anything else has in years. He now uses a long walk, dance class, or opportunity to garden outside to get himself out of almost any funk—and this, paired with the Better Brain Diet, has allowed Frank to lose one hundred pounds, shaving eighteen inches off his waistline.

KRISTEN'S TIP: Discovering a form of exercise you enjoy is key to getting active and making it a lifelong habit. After Frank found walking and dancing, he was able to change his brain and body in amazing ways.

Exercise During the Day Can Help Your Brain Sleep Better at Night

If you're tossing and turning at night and not exercising during the day, don't blame your sleep problems on a stressful work project. Study after study shows physical activity helps you fall asleep faster, wake up less frequently, and get up the next morning feeling more refreshed.[28] Even just ten minutes of daily exercise is enough to boost sleep quantity and quality.[29] Working out also decreases the risk of developing disorders like insomnia, sleep apnea, and restless leg syndrome.

BEST EXERCISES TO IMPROVE SLEEP: Almost any kind of exercise helps increase sleep quality and quantity, according to the National Sleep Foundation. While most of the research has been conducted on aerobic exercise like walking, running, and cycling, studies on yoga have shown it to be effective at helping people sleep more soundly, due to its calming and grounding nature.[30]

How to Lose Weight and Pump Up Your Brain at the Same Time

As anyone with a television or computer knows, exercise helps you lose weight. While many debate the best form of physical activity for fat loss, the science is conclusive that moving your body will help you drop pounds.

While the brain itself doesn't contain fat cells, carrying excess fat elsewhere affects the mind profoundly. Researchers now believe fat cells release harmful toxins that can permeate the

The Best Time for Your Brain to Exercise

From a physiological standpoint, morning exercise is more advantageous because it resets the body's circadian rhythm, or sleep-wake cycle. That's because exercise raises our core temperature, signaling it's time to wake up.[31] Our natural cortisol levels are also higher in the morning, and since exercise elevates the hormone, working out soon after you wake up helps keep your cortisol in sync. Finally, if you exercise outside in the morning, you're exposed to sunlight, even on a cloudy day, which suppresses melatonin and stimulates mood-boosting serotonin. Consequently, your body learns to produce melatonin earlier in the evening, helping you fall asleep faster.[32]

Research also shows exercising before work or school improves focus and the brain's ability to ideate, create, and learn. Despite this research, if you love an evening run, a postwork dance class, or a nightly trip to the gym, keep doing it. Science aside, the best time to exercise is whenever you're most likely to do it. Just be sure to finish any physical activity at least two hours before bed to give your brain enough time to clear cortisol and the other energizing chemicals produced by your workout.

blood-brain barrier—the boundary that separates the brain's blood vessels from its tissues and cells. That can amount to lots of toxins if you're carrying too much body fat.[33]

Once inside the brain, toxins from fat cells can cause a takeover. They invade the hippocampus and interfere with its function, causing cognitive synapses to misfire and malfunction. The result is impaired memory, slower learning, and overall cognitive decline.[34]

Personally, I've conducted research with colleagues that shows adults who are overweight or obese have lower blood flow to their prefrontal cortex, the area of the brain responsible for high-level cognitive thinking.[35] In our NFL study, we found that those who were overweight had decreased blood flow to two areas of the brain, the prefrontal cortex and temporal lobes, which negatively impacted the player's mood, memory, and overall cognitive performance.[36]

Losing weight is obviously the first solution. But simply starting to move more if you are overweight will help fight the effects. Scientists have found that mice that exercise are able to reverse fat-inflicted damage on their brains and even normalize hippocampal function. Active animals also perform better on cognitive tests than sedentary ones, even when they weigh the same amount.[37] Similarly, a study in humans found that overweight and obese people who exercise for two months are able to boost cerebral circulation and undo some toxic effects in the brain.[38]

If you're thinking about losing weight without exercise, you'll still reap plenty of brain benefits, but not as many as if you combine it with exercise. Similarly, a published study showed that thin people who are sedentary have lower cognitive function than those who are heavier but also physically fit.[39]

BEST EXERCISES TO LOSE WEIGHT AND BOOST BRAIN HEALTH: A recent study conducted on more than eighteen thousand people found that those who jogged were able to lose the most weight and keep it off over time.[40]

What's so great about running? It's a full-body workout, keeps your heart rate in the fat-burning zone, and is super accessible. All you need is a pair of shoes—no gym, extra equipment, or training partners—and you're off, anywhere, anytime.

Short bursts of high-intensity exercise like sprints can help torch fat more quickly, according to research. Known as high-intensity interval training (HIIT), studies show fast-paced pick-ups can burn more fat than the same activity (e.g., running, cycling, swimming, power walking) done at a slower pace for a longer duration.[41]

Know Someone with a Cognitive Disorder? Here's How to Help

Exercise isn't usually top of mind for those suffering from neurodegenerative disorders. But encouraging someone with dementia to be more active can dramatically improve his or her cognitive health and overall quality of life.

When my father was struggling with tremors and balance problems associated with Parkinson's, he started going to the gym to ride a stationary bike for thirty minutes to an hour three to four times per week. This was in addition to the physical therapy, stretching, and lightweight lifting he already did. The exercise helped his body to relax, which reduced his tremors, allowing him to move more easily, despite the muscle rigidity and balance problems he faced. Mentally, the cycling made him sharper and more able to focus, as he started doing his *Chicago Tribune* crossword puzzles quicker and with more zeal.

True to my father's story, brain-imaging studies show cycling at a certain rate improves functional connectivity in brain networks in those with Parkinson's and may be a safe, low-cost way to help manage motor symptoms associated with the disease.[42] Similar research has also found people with early Alzheimer's or other forms of mild cognitive impairment can help increase their brain volume and maintain cognitive function by adding weekly aerobic exercise to their routine.[43]

5
THE SUPPLEMENT OFFENSIVE

In my late teens, I spent a summer working as a front-desk attendant at a fitness club outside of Barrington Hills, the suburb of Chicago where I grew up. The club was popular with competitive bodybuilders, who used an array of protein shakes, amino acid powders, and other nutritional supplements to build lean muscle and take weight off quickly.

While I certainly wasn't a bodybuilder, I was modeling and subsisting on a diet of lean protein, vegetables, and MET-Rx shakes. Those were especially trendy at the time, and while my shake habit was short-lived, it taught me that you could transform your body quite effectively by supplementing with the right products.

During graduate school, I became curious how micronutrients—the vitamins, minerals, antioxidants, amino acids, and other essential nutrients the body needs—could help people transform not just bodies but also manage symptoms associated with brain disorders. In particular, I was interested in Parkinson's disease—my area of the research—and whether micronutrients could counter the oxidative stress in the brain that accelerates disease progression. I started attending support groups for people with Parkinson's. Hearing patients' stories and seeing how helpless many felt in the face of the disease inspired me to try to find a way to use my research to give them back some control over their bodies and brains—empowering them to manage the symptoms they felt were managing them. During this time, I never imagined my work would benefit my father a decade later when he, too, was diagnosed with Parkinson's.

It wasn't until 2009, when my colleagues and I conducted a clinical trial with pro players, that I truly understood the impact supplements can have on brain function. When we first saw the players' scans, the majority had mild brain damage and cognitive impairment. Only a few were getting enough blood to the brain, and many had obvious impairments in cognitive function. By putting them on a daily regimen of specific supplements, we hoped that would be able to increase perfusion—a fancy word for improving blood flow to the brain—and reverse some of the damage.

After six months, the players' follow-up brain scans told a completely different story than their baseline images, showing increased blood flow in areas important for executive function, memory, vision, and coordination.[1] From a neurocognitive perspective, here's what our final data showed:

- Nearly half demonstrated a greater than 50 percent increase in cognitive function and proficiency
- 69 percent showed improved memory
- 53 percent showed improved attention
- 38 percent reported significant improvements in mood

- 38 percent reported significant improvements in motivation
- 25 percent reported significant improvements in sleep

These results were exciting. But they became life-changing for us at the clinic after we replicated the effects, improving cerebral blood flow and neuropsychological function using the same supplements at a lower dosage in people who didn't have the same extent of cognitive damage.[2] Over and over again, through the data and our clinical work, we saw that people can change their brain health—and even reverse damage—simply by adding supplements that support brain-smart nutrient choices.

In an ideal world, I'd recommend anyone concerned with cognitive health and performance to consider taking most of the supplements detailed in this chapter. All have been shown by science to support optimal cognitive health and function, as long as you choose a reputable brand and follow the dosing recommendations.

I also understand not everyone has the time, money, or faith (yet) for a daily supplement protocol of over a dozen pills and powders. That's why I've outlined three tiers here that will allow you to choose a level of commitment based on your goals and personal needs:

- Starting Lineup, or basic nutrients I believe everyone should take for basic brain health
- All-Star Team, or those nutrients you can add to the Starting Lineup if you want to do everything possible to boost your cognitive health
- Injured Reserve, or supplements I recommend for those with a concussion, mild traumatic brain injury, or a cognitive disorder

Starting Lineup: The Six Supplements to Change Your Brain

If you take these six supplements daily for at least three months, you can start to change your brain. I'm not saying this because I own stock in supplement companies or believe swallowing every pill down the vitamin aisle will miraculously cure whatever ails you. Quite the opposite—I know some supplements have no science to show they benefit the brain or body in any way. I was even skeptical of supplements myself until I saw during our clinical trials the impact that they can have on brain health and function.

That said, I encourage you to approach this list with an open mind. From a neurocognitive perspective, taking a curated, science-backed set of supplements is one of the best investments you can make in your cognitive health.

Omega-3 Fatty Acids

WHY: Omega-3 fatty acids make up every cell membrane in our body and are needed by neurons to operate on a basic level. These fats also fight oxidative stress and lower inflammation, reducing the risk of cancer, heart disease, depression, arthritis, ADHD, and a host of other physical and mental conditions.

But not all types of omega-3s wield wonders in the brain and body. The omega-3s responsible for most of these benefits are what are known as the marine omega-3s: docosahexaenoic acid (DHA) and eicosapentaenoic acid (EPA). DHA and EPA are found only in seafood and edible algae like seaweed and spirulina—foods most Americans infrequently consume. In fact, a recent survey found almost half of all American's eat little to no seafood.[3] That's a problem because our body can't manufacture DHA or EPA on its own. For these reasons, sources estimate up to 90 percent of all Americans have dangerously low levels of marine omega-3s.

Compounding the problem, we eat too much omega-6, a fat found in vegetable oils, nuts, seeds, fatty meat, and processed foods. The body needs some omega-6 fat for cell growth and function, but consuming too much alters the body's delicate balance of omega-3s to omega-6s, which, in turn, increases inflammation and the risk for dementia, heart disease, stroke, and other ailments. Our consumption of omega-6, thereby, raises the body's need for more omega-3.

Science shows that taking an omega-3 supplement rich in DHA and EPA on a daily basis can increase cerebral circulation, help support new neuron growth, improve cognition, and reduce inflammation. Supplemental DHA can also increase feel-good chemicals like serotonin while reducing the risk of depression, anxiety, diabetes, weight gain, high blood pressure, cancer, heart diseases, and neurodegenerative disorders like Alzheimer's.

TIPS: Look for an omega-3 supplement that includes a seal showing it's been tested for heavy metals like mercury. I prefer to take enterically coated fish-oil supplements, which means they help prevent "fish burps," or any unpleasant aftertaste from the supplement. Since omega-3 supplements can thin the blood, consult your doctor before taking omega-3s if you're on blood-thinning meds like Warfarin.

Multivitamin

WHY: A multivitamin for the brain is the dietary equivalent of a garage for your car. Do you need it? No, you can still own a vehicle without one. But having a garage will ensure your car lasts longer, functions better, and remains protected from the elements. A multivitamin works the same way for the brain.

During our clinical trial, we saw that NFL players who took a high-potency multivitamin were able to plug the nutritional deficiencies that can impede cerebral circulation and cognitive function. In healthy adults, studies show supplementing daily with a multivitamin can delay the onset of age-related decline by as much as five years. Even older adults with mild brain impairment who take a daily multivitamin can significantly increase cognitive function.[4]

Isn't a balanced diet the best way to get all essential nutrients? It absolutely is, but very few of us eat a balanced diet. Approximately 90 percent of all Americans don't get enough vitamins D or E from food, while half don't consume enough vitamin A and magnesium. Just under 50 percent of all Americans get enough calcium and vitamin C from food. Zinc, folate, and many B vitamins are also lacking from the standard American diet.[5]

Even those of us who do follow a healthy diet can benefit from taking a multivitamin. That's because even the healthiest diet in the world can't counter absorption issues from prescription medications, gastrointestinal problems, and certain lifestyle choices, like drinking alcohol or even working out too much in the gym.

TIPS: I recommend taking a high-potency multivitamin made from whole-food sources like fruits and vegetables. Avoid gummy, chewable, or other candy-like multivitamins, which contain sugar, food dyes, and other unhealthy additives.

Probiotics

WHY: Probiotics are the healthy bacteria that live inside our gut. These living microorganisms are also found in foods like yogurt, buttermilk, miso, kimchi, tempeh, unpasteurized sauerkraut, and other fermented and cultured products. Probiotics, which means "for life" in Latin and Greek, balance out the unfriendly bacteria in the gut. While you can't escape bad bacteria, the more probiotics you consume, the healthier your overall microbiome—the trillions of bacteria, fungi, and other microorganisms in the human body—will be and, as a result, your brain and the rest of your body will be healthier as well.

What do probiotics do? The beneficial bacteria help us digest food, boost immune function, and ensure skin stays smooth and healthy. Probiotics also are critical to nutrient absorption, and if we don't have enough, dietary supplements aren't as effective, since the body may not be able to assimilate them.

From a cognitive perspective, probiotics help the GI tract produce feel-good neurotransmitters, including dopamine, GABA, and serotonin, 90 percent of which is created in the digestive tract. For this reason, probiotic supplements have been shown to boost mood and lower anxiety and stress.

Probiotic supplements also help preserve and improve cognitive function, according to research.[6]

TIPS: Look for a high-potency probiotic that includes a range of bacteria species, otherwise known as strains. Tests show some probiotic supplements don't include the number of live bacteria listed on the label—be sure to choose a trusted brand that's been tested for potency or includes a seal of quality. A good way to weed out the imposters is to choose a supplement that includes an expiration date: Probiotics are living organisms, so they will expire if left too long on the shelf.

Vitamin D

WHY: Known as the "sunshine vitamin," Vitamin D is a fat-soluble nutrient that helps bones and teeth absorb calcium and is essential for proper immune function and cell growth. Low levels of vitamin D can create inflammation and insulin resistance, leading to weight gain and diabetes. The vitamin also affects how some genes are expressed, particularly those involved in cancer development. Inadequate amounts of vitamin D are also associated with higher cancer risk.

In the brain, vitamin D influences neuron function and helps buffer calcium levels that otherwise can cause depression and other mood disorders.[7] The nutrient also aids in clearing the buildup of amyloid plaque, which is associated with dementia and Alzheimer's.[8] Without enough vitamin D, inflammation and insulin resistance can increase, leading to cognitive dysfunction and metabolic disorders.

The optimal way to get vitamin D is through food, but the products highest in the nutrient—cod liver oil, salmon, tuna, beef liver—are hardly standard fare. Egg yolks also contain some vitamin D, along with fortified dairy products. Even still, the only source that supplies enough vitamin D in a single serving is cod liver oil.[9]

For all these reasons, a supplement is crucial to helping regulate mood, fight depression, improve memory, and prevent neurodegenerative disorders. Supplementing with vitamin D may also reduce your risk of cancer, diabetes, osteoporosis, and other ailments.

TIPS: Look for a supplement that contains vitamin D3, the biologically active form of the nutrient.

Liquid Trace Minerals Formula

WHY: When you think about the minerals we need for good health, calcium, magnesium, sodium, potassium, and other macrominerals come to mind. But there's a whole other set of minerals known as microminerals, or trace minerals, that our body and brain need just as much as these major macrominerals. Microminerals include boron, chromium, copper, germanium, iodine, iron, manganese, molybdenum, selenium, silicon, sulfur, vanadium, and zinc.

While each micromineral has a different function, trace minerals in general help us manufacture enzymes, hormones, and cells critical to cognitive function. Some trace minerals also serve as antioxidants in the body, helping lower inflammation. Others are needed to make neurotransmitters or to clear the body and brain of toxins.

Being too low in any trace mineral can cause serious side effects. In the brain, a deficiency can manifest in many ways, depending on the nutrient, but in general, suboptimal levels can trigger mood problems and reduce mental capacity. Having low of levels of zinc can slow executive function, impair memory, and speed age-related cognitive decline.[10] Having too little selenium, on the other hand, can interfere with neuron function and learning.[11] Being too low in chromium can cause blood sugar imbalances, neurotransmitter deficiencies, depression, and weight gain.

The best way to make sure you're getting enough trace minerals is to consume a variety of vegetables, fruits, and other fresh plants. Unfortunately, most Americans don't eat enough plants—and if they do, they aren't eating the variety we need for adequate intake. Trace minerals have also declined over the years in soil and rock due to high farming yields, making plants less nutrient-dense than they were decades ago. If you have malabsorption problems, you may need to supplement with trace minerals, no matter how healthy and diverse your diet is.

While you can get a blood test to help determine adequate levels of trace minerals—and I encourage you to do this if you have the opportunity—taking a trace mineral supplement is simple insurance for your body and brain that you're getting enough of these vital nutrients.

TIPS: Look for trace mineral supplements in liquid form, which is more easily absorbed than minerals in tablet or capsule form.

Curcumin

WHY: If you've ever eaten Indian food, you're likely familiar with curcumin, the active compound in turmeric that gives the spice its golden hue. Mounting research proves curcumin is one of the most potent supplements for brain health, capable of lowering inflammation and improving cognitive function.

Curcumin, which has been used as medicine in India for centuries, is a powerful antioxidant, neutralizing free radicals that can create oxidative stress and damage brain cells. The compound also improves the body's own antioxidant defenses, helping you better fight stress in the future.[12]

Curcumin is also a potent anti-inflammatory, working to block molecules that activate genes associated with inflammation.[13] What's more, curcumin stimulates brain-derived neurotrophic factor (BDNF), the protein responsible for the survival, growth, and maintenance of healthy neurons.[14] Optimized levels of BDNF in key brain regions related to mood has antidepressant effects and plays a preventative role in reducing the risk of Alzheimer's and other neurodegenerative diseases.

Cognitive tests show curcumin supplements can also improve memory and attention.[15] Outside the brain, curcumin has been shown to prevent and even treat cancer, joint pain, and heart disease.[16]

Unfortunately, you can't just season food with turmeric and reap benefits: The spice contains too little curcumin—only around 3 percent by weight.[17] Research demonstrating the efficacy of curcumin on brain health parameters has been primarily conducted on supplements.

TIPS: Be sure to purchase a curcumin supplement that contains piperine, a black pepper extract, which helps the body absorb the compound into the bloodstream.[18] If you're on blood-thinning medications like Warfarin, supplementing with curcumin is not recommended, since the nutrient has anticoagulant properties.

Astrid's Story

USING SUPPLEMENTS FOR THAT EXTRA EDGE IN YOUR BRAIN GAME

Astrid, fifty-nine, took care of her in-laws when they suffered with dementia and Alzheimer's. After both passed away, she felt mentally and emotionally drained by the pressure to remain strong for her husband, three daughters, and grandchild.

An active golfer for more than three decades, Astrid had recently started to lose her mental edge around the thirteenth or fourteenth hole. She knew there had to be a way to improve her focus, but she was tired of throwing money away on supplements and treatments that didn't work. She told me she reached out to me because she wanted someone who knew which supplements had the best clinical evidence behind them for cognitive health.

Astrid's brain scan looked beautiful, showing no early signs of dementia or other cognitive issues. This was a relief to her, obviously, and a full challenge to me to improve a perfectly healthy brain through a dedicated regimen of supplements.

Astrid had been taking a multivitamin for years, but she never believed the supplement supported her cognitively. I suggested she stay on it while adding omega-3, curcumin, spirulina, and vitamins C and D supplements.

Within three months, Astrid noticed she had more mental focus and agility, and her golf game had improved as a result. At home, when her grandchild had a meltdown, Astrid felt like she had more patience and mental resilience to handle it.

Astrid began to experiment with the timing of her supplements, taking certain ones before her golf game to see if that made a difference. Through trial and error, she found that if she took spirulina before hitting the links, she had more focus, so she started consuming the nutrient about an hour before her games.

The few times Astrid fell off her supplement routine, she told me she noticed the effects. Once, during a trip out of town, she forgot to bring her supplements, which led to brain fog and poor performance the four times she played golf.

Today, Astrid still takes spirulina, omega-3s, vitamin D, and curcumin every day. The regimen helps her manage her expanded family, which now includes six grandchildren and two dogs. Her husband, a former longtime supplement skeptic, now follows her protocol after seeing the effects supplements have had on his wife.

KRISTEN'S TIP: Taking science-backed supplements specific to cognitive function can make a big impact on many areas of your life, improving your ability to manage your family or career, pursue your favorite hobby, or become better at an activity or skill.

The All-Star Team

The Starting Lineup is a great place to begin your supplement protocol. But if you want to do more for your brain, consider adding the following supplements to that list. All five listed here are well-researched nutrients that have proven effects on cognitive function.

B Complex

WHY: This umbrella term for the family of B vitamins includes eight essential nutrients we need on a daily basis. B complex is essential to nearly every physical operation, including the conversion of food into energy, the creation of new blood cells, proper cardiovascular health, adequate energy levels, hormone production, cholesterol balance, metabolic function, and muscle health.

B complex is critical to our central nervous system. All eight B vitamins can cross the blood-brain barrier, where each plays a vital role in cognitive function. The most important Bs for the brain are B_6, B_{12}, and B_9, also called folate (or folic acid in supplemental form).

If you've ever been anemic or iron deficient, you're likely familiar with folate, which helps the body manufacture red blood cells, white blood cells, neurotransmitters, and DNA. Folate also helps break down homocysteine, the amino acid linked to Alzheimer's risk. Low folate can also speed brain aging and trigger cognitive dysfunction. Without enough folate, your mood can also suffer—many deficient in B_9 have clinical depression or other psychiatric disorders.[19]

As to B_{12}, not getting enough of this nutrient can cause memory problems, kill neurons, and speed age-related brain shrinkage.[20] Psychologists and psychiatrists often test patients' B_{12} levels because inadequate intake can double the risk of depression.[21] There's also strong evidence showing low levels of B_{12} can increase Alzheimer's risk. Studies show being too low in this nutrient can also trigger dementia-like symptoms.[22]

Like folate and B_{12}, B_6 helps the body produce neurotransmitters like serotonin. Inadequate B_6 can slow memory, impair concentration, and significantly increase the likelihood of mood disorders.

You don't need a clinical deficiency in these B vitamins to

develop indications of cognitive dysfunction and mood issues. B_6 and folate are found in many foods, but we don't absorb the nutrients as efficiently with age. Being overweight, drinking alcohol, and taking some medications also deplete B vitamins. If you're vegetarian or vegan, it's also easy to become deficient in B_{12}, as the nutrient is found primarily in animal products.

TIPS: Look for a complex that contains all eight B vitamins, as the nutrients work better in concert together. Also choose one that contains the natural bioidentical form of B_{12} such as methylcobalamin, as opposed to a synthetic form such as cyanocobalamin, which isn't as bioavailable.[23]

Vitamin C

WHY: Many load up on vitamin C when they have a cold or other respiratory illness. But there's a host of reasons why you should supplement year-round with this nutrient, which is one of nature's most powerful antioxidants. In fact, research shows that supplementing with vitamin C for five years daily can increase blood serum levels by as much as 30 percent, boosting your antioxidant levels.[24] This helps fight free radicals, slashing oxidative stress and the kind of inflammation associated with dementia.

Increasing your intake of vitamin C through food and supplements has been shown to improve memory and executive function.[25] Older people who regularly take vitamin C also have fewer symptoms of cognitive decline.

Prioritize getting vitamin C from foods like strawberries, oranges, lemons, asparagus, avocados, broccoli, and other fruits and vegetables. But since it's difficult to obtain enough C through food alone, especially in the higher levels shown by research to be more beneficial to the brain, add a daily supplement to your healthy diet.

TIPS: Your body can absorb only so much vitamin C at one

time, so split your dose in two, taking half the recommended amount in the morning and the other half at night.[26] If you're on blood-thinning meds like Warfarin, consult with a doctor before supplementing with vitamin C.

Magnesium

WHY: Most Americans don't consume enough magnesium, which is critical to every cell in the body and brain. Magnesium aids in more than three hundred biochemical reactions, including functions that affect stress, neurotransmitter production, muscle relaxation, and hydration.

Without enough magnesium, your risk of depression skyrockets, along with feelings of anxiety, aggression, irritability, and brain fog. Low levels of the mineral also prevent your brain from working efficiently. A magnesium supplement, on the other hand, has been shown to boost learning and memory while preventing age-related cognitive decline.[27] People with mild brain damage can also improve cognitive function by taking magnesium.[28]

Many foods contain magnesium, including cashews, brown rice, kale, spinach, almonds, black beans, quinoa, and sunflower seeds, but it's difficult to obtain enough of the nutrient through food alone. Approximately half of all Americans don't meet the basic daily requirement for magnesium, and many more have suboptimal levels.

TIPS: Look for the organic forms of magnesium citrate or amino acid chelates; both formulas are more easily absorbed.[29]

Spirulina

WHY: We now know the marine omega-3s DHA and EPA are critical to cognitive health, and it's difficult to get enough of these essential fats through food alone. Spirulina is a type of blue-green algae rich in DHA and EPA that helps increase omega-3 consumption. The algae, available in powder or tablet form, also contains many of the nutrients recommended in this chapter, like magnesium, zinc, B_{12}, B_6, and folic acid. Spirulina also has all nine essential amino acids, including tryptophan, which your body needs the most to manufacture serotonin. Studies show spirulina can help maintain healthy gut bacteria, lower blood sugar, aid in weight loss, and destroy free radicals.

TIPS: You can find spirulina at most natural-food stores. If you opt for a powder supplement, stir it into water or smoothies.

Coenzyme Q10

WHY: Coenzyme Q10, or CoQ10, is a nutrient that acts as an antioxidant and is needed by cell mitochondria to generate energy and feed the body and brain. CoQ10 has been shown to fight oxidative stress in powerful ways, reducing the risk of heart disease, cancer, and neurodegenerative disorders like Alzheimer's and Parkinson's.

Our brain consumes 20 percent of our body's oxygen, and CoQ10 helps the brain to maintain its energy requirements by keeping it well fueled and high functioning. Animal research shows CoQ10 improves learning, memory, and overall cognitive performance,[30] and studies are now investigating its effect on the brain in healthy older populations.[31]

CoQ10 is found in foods like organ meat, animal meat, and fatty fish, but the most promising research for cognitive health has been conducted on high-dose CoQ10 supplements. Our body can't store the antioxidant, which makes supplementation advantageous, and CoQ10 levels appear to decline with age.

TIPS: Choose a product made with ubiquinol, which is more bioavailable than ubiquinone, the common form of CoQ10. Taking the nutrient with food increases absorption.[32]

The Injured Reserve

If you've sustained a traumatic brain injury or have early dementia, you'll want to add these seven supplements to your arsenal to support restoration of brain function and prevent further impairment.

Phosphatidylserine

WHY: Phosphatidylserine (PS) is a fatty component of every cell membrane in the body and brain. The material is responsible for healthy nerve function and helps make up myelin—the fatty sheath that surrounds nerve cells and helps our brain to send messages faster and more efficiently. PS, as a part of the cell membrane, allows for the delivery of nutrients and the removal of waste from neurons.

PS levels in the brain decline with age, which slows cell signaling and may impact our memory, mood, and executive functions. Taking a PS supplement, however, has been shown to stop and even reverse age-related nutrient decline.[33] In particular, supplementing with PS can improve the formation and consolidation of memories, the ability to learn new information, concentration, communication, and problem solving.[34] People with Alzheimer's who take the supplement for six to twelve weeks may also be able to reduce some symptoms of the disease, according to research.[35]

PS has also been shown to curb depression and regulate mood, even when taken for only a handful of weeks.[36] The supplement shows promising research for ADHD, countering the hyperactivity, impulsive behavior, and poor mood associated with the condition.[37]

PS is found in some foods, primarily soy, but also egg yolks, animal liver, and white beans. But to get enough for healthy brain function, you should consider adding it to your supplement regimen.

TIPS: Look for food-based PS supplements made from soy or cabbage.

N-acetylcysteine

WHY: N-acetylcysteine, or NAC, is the supplemental form of the amino acid cysteine, which helps the body manufacture proteins like collagen. The body needs NAC to create the antioxidant glutathione, which helps neutralize free radicals. NAC also regulates the neurotransmitter glutamate, which helps send signals between neurons and is generally acknowledged by researchers as the most important neurotransmitter for healthy brain function.[38]

For those with a mild traumatic brain injury or early dementia, NAC significantly lowers levels of homocysteine,[39] which, when elevated, can lead to cognitive damage, dysfunction, and Alzheimer's disease. The nutrient also binds to heavy metals like lead, mercury, and other pollutants that can accumulate in brain cells. Finally, NAC is a vasodilator, relaxing blood vessels and helping speed oxygen delivery to the brain.

TIPS: Talk with your doctor before supplementing with NAC if you're already taking blood thinners like Warfarin or if you have asthma.

Acetyl-L-carnitine

WHY: Acetyl-L-carnitine, or ALC, is the supplemental form of the amino acid carnitine, which helps brain cells produce energy. Similar to NAC, ALC is also an antioxidant that fights free radicals and inflammation. The nutrient also helps repair neuron damage, making it essential for those who've suffered a concussion or other type of mild traumatic brain injury. Studies also show people who supplement with high doses of ALC improve their reaction time, memory, and cognitive function. The nutrient also appears to prevent age-related brain decline, along with the cognitive dysfunction that can stem from dementia.[40]

For those with mood problems like depression, ALC has been shown to increase levels of the energizing and feel-good neurotransmitters, noradrenaline and serotonin.[41] Supplementing with ALC may even fight mild depression as effectively as prescription drugs.[42]

TIPS: Talk with your doctor before supplementing with ALC if you have a thyroid problem or are taking blood-thinning meds like Warfarin.

Huperzine A

WHY: This natural chemical compound, derived from the Chinese club moss plant, is accumulating more promising research as a potential treatment for Alzheimer's disease—the extract is already licensed as an anti-Alzheimer's drug in China.[43] Research shows taking the supplement over time can improve memory and mental function, both in healthy individuals and those with Alzheimer's and other forms of dementia. Huperzine A works by increasing production of the neurotransmitter acetylcholine, which has been shown to improve cognitive function, attention, and alertness.

TIPS: Consult with your doctor before taking huperzine A if you're on beta-blockers or anticonvulsive meds or have already been diagnosed with Alzheimer's or another form of dementia.

Vinpocetine

WHY: Like huperzine A, vinpocetine comes from a plant—in this instance, the seeds of the periwinkle plant. The extract is considered a supplement in the United States, but in parts of Japan, Europe, Mexico, and Russia vinpocetine is prescribed as a drug to help boost brain blood flow, neuronal metabolism, and overall cognitive function in those with stroke or other cerebrovascular disorders.[44]

For those with brain injuries or other cognitive problems, vinpocetine can increase glucose and oxygen consumption in the brain. The extract is also a vasodilator, working to open blood vessels and speed cerebral circulation. Scientists who have conducted studies on the extract recommend vinpocetine to those with mild cognitive impairment.[45]

TIPS: Consult your doctor before taking vinpocetine if you're already taking blood pressure or blood-thinning drugs.

Ginkgo biloba

WHY: This supplement, derived from the leaves of the ginkgo biloba tree, has been prized as a powerful antidote in Chinese medicine for centuries. Today, the extract is used as a drug in Europe to treat early Alzheimer's and dementia.[46] Research shows the nutrient had modest effects on improving symptoms and cerebral perfusion deficits associated with Alzheimer's.[47] Ginkgo may also support memory and cognition in healthy people.[48]

One reason gingko biloba has these effects is because it can improve circulation, widening blood vessels and making blood less sticky, which increases blood flow to the brain. The nutrient is also recommended by many integrative health practitioners for general circulation problems. It is also a potent antioxidant, helping to destroy free radicals before they harm cells.

TIPS: Since gingko biloba has powerful effects on circulation, consult with a physician before supplementing if you are already on any medication that thins blood, including Warfarin, aspirin, antiplatelet drugs, diabetes medications, NSAID painkillers, anticonvulsants, antidepressants, and liver drugs.

Alpha-lipoic acid

WHY: Alpha-lipoic acid is an antioxidant that helps protect brain cells against oxidative stress. Unlike most antioxidants, which are either fat-soluble or water-soluble, alpha-lipoic acid is both, allowing it to cross the blood-brain barrier more easily and work effectively in different types of tissue.

Alpha-lipoic acid has been shown to clear brain cells of heavy metals and fight against the age-related slowdown of neurotransmitter production, which can lead to mood problems and memory loss.[49] Some with Alzheimer's and other forms of dementia who supplement with alpha-lipoic acid have been able to prevent memory loss and improve overall cognition.

Alpha-lipoic acid is found in low amounts in foods like red meat, spinach, broccoli, potatoes, and yeast, but supplements are the only way to get optimal amounts of this potent antioxidant.

TIPS: If you have diabetes or other blood sugar issues, consult with your doctor before taking alpha-lipoic acid as it can lower blood sugar levels. Alpha-lipoic acid may also interfere with some chemotherapy and thyroid medications.

The NFL Story

SAVING THE BRAIN WITH SUPPLEMENTS

John had played football his entire life, having been an All-American in high school before spending four years in the NFL as an offensive lineman. Not surprisingly, he had grave concerns about his brain health after years of hard hits. John, like many players, was worried about chronic traumatic encephalopathy (CTE), and he told me he didn't want to sit on the sidelines only to start showing symptoms associated with the degenerative brain disease five, ten, or twenty-five years later.

When I started working with John, he was already taking supplements, including a daily multivitamin and vitamins C, D, and E, along with omega-3s sporadically. But he told me he had only adopted the protocol to support his body and wanted to do more for his brain.

After the success of our trial with NFL players, I recommended John take a high-potency multivitamin and more omega-3 fatty acids on a regular basis, along with all the supplements we'd given the players (listed in the Injured Reserve protocol on page 115), including phosphatidylserine, N-acetylcysteine, acetyl-L-carnitine, huperzine A, vinpocetine, gingko biloba, and alpha-lipoic acid.

John took the supplements daily for more than a year while following the Better Brain Diet and continuing to lift weights, run, and do some high-intensity interval training. Despite taking eight to fifteen pills at one time, John had no problem with the supplement regimen, always remembering to take them in the morning with breakfast and at night with dinner, so they were as much a part of his meal as a fork and knife.

After six months, John told me he had more mental clarity and focus, and that his memory seemed sharper. He couldn't believe that no one had taught him or his teammates that certain supplements, along with diet, exercise, and stress reduction, could have such an impact on his brain health.

After eighteen months on his supplement protocol, John reached out to tell me the regimen had made a noticeable difference to his brain health. His cognitive function had improved significantly, and he felt sharper and more focused. He had even started experimenting with other supplements that were supported by science, like the root rhodiola, shown to help control anxiety.

KRISTEN'S TIP: Rhodiola is an herb that's been used for centuries to help ease anxiety and fatigue. I recommend it to reduce stress, boost mood, and enhance energy. Since it's an antioxidant, rhodiola also helps protect the brain from the effects of stress.

Eight Secrets to Getting the Most Out of Your Supplements

1. **COMMIT TO THE INVESTMENT.** Supplements aren't cheap. While you can find inexpensive options, adopting a comprehensive protocol will inevitably cost money. Think about it as an investment—spending a little money now could save you thousands in medical costs down the road.

2. **SHOP ONLINE.** I shop online for my supplements, where it's easier to find the brand, specifications, dosing, and price I want. Many online retailers also offer discounts if you subscribe to refill a supplement every month, three months, or six months.

3. **LOOK FOR CERTIFICATIONS.** Supplements are a multi-billion-dollar industry, unregulated by the FDA. To protect yourself, look for a blue-and-yellow seal from the United States Pharmacopeia (USP), or a symbol from NSF International or ConsumerLab.com. All three organizations independently test supplements for ingredient quality and potency. You should also try to find a product made in a Good Manufacturing Processes (GMP) facility, which helps ensure strength, composition, quality, and purity of the product.

4. **RESEARCH FOR THE RIGHT BRAND.** Research before you buy and find out where a supplement sources its ingredients—the best brands will reveal this information. Personally, I like to buy from companies that use only pure nutrients or compounds scientifically tested by independent laboratories. A slightly deeper online search is usually all it takes to uncover which brands are backed by good science.

5. **AVOID ADDED INGREDIENTS.** If you're taking a supplement every day for the sole purpose of augmenting your health,

make sure it doesn't include anything that could potentially harm your body or brain at the same time. This includes artificial colors, flavors, sugars, and magnesium stearate, a compound often added to supplements that may impair immune system function. A pro tip is that supplements containing gluten, soy, corn, or dairy are often low-quality.

6. **STORE YOUR SUPPLEMENTS THE RIGHT WAY.** Keep your supplements in a cool, dry place, away from direct sunlight and any sources of heat, air conditioning, and extreme cold. Be sure to screw on lids tightly, as many supplements like omega-3s can go bad if exposed to too much oxygen. Finally, pay attention to the expiration date. Supplements can and do lose potency over time.

7. **GIVE IT THREE MONTHS.** Remember that supplements aren't like prescription drugs or surgery, and they may take several months of daily use before improving symptoms or affecting cognitive function. Don't give up.

8. **ALWAYS TALK WITH YOUR DOCTOR.** While I've included instances throughout this chapter when you'll want to consult with a doctor, everyone should speak with a health care practitioner before starting a new supplement routine. Be sure to discuss existing medical concerns and specific health goals, along with the other dietary supplements or over-the-counter drugs you already take, since some can interfere with the nutrients you want to add. Always talk with a doctor about your supplement routine before surgery.

6

THE HYDRATION OFFENSIVE

Drinking to biohack your brain goes far beyond simply consuming more water. Just like a high-performance sports car requires high-octane fuel to run efficiently, your brain needs clean water to attain peak cognitive function. Believe me, the side effects of poor hydration habits can be brutal; in fact, my own story may make you think twice about how you hydrate.

While I was finishing my Ph.D., I started to give presentations on my doctoral research. At the time, I didn't drink much water; I assumed I was plenty hydrated because I drank several green juices per day and ate a ton of fruits and veggies. One of my first presentations was in Los Angeles at Cedars-Sinai Medical Center, which has been consistently ranked as one of the best hospitals in the country, and is definitely one of the most intimidating to speak at.

While I had given many presentations before, this was the first time I passed out in front of an audience. I was talking about my clinical research in front of the hospital's neurology and neurosurgery department when I felt light-headed and started seeing stars. I thought to myself, *Oh my god, I'm going to faint*. And I did, only ten minutes into the talk! Fortunately, I fell backward into a chair that just happened to be behind me. The next thing I remember, I was coming to and asking if I could continue my talk (they politely told me it was best to reschedule).

Afterward, I was curious why it had happened. Was it nerves? I had given numerous presentations before, so I brushed it off as an odd occurrence.

The next time I fainted, though, the stakes were higher. I was presenting at the National Institutes of Health in Bethesda, Maryland, where I had been awarded a research grant. I was speaking in front of a group of people about my research when I simply collapsed to the ground. I woke up on the floor with several doctors hovering over me. I remembered one looked at me and said, *Well, if you're going to faint anywhere, you just did so at the best facility in the world.*

Now I knew I had a problem. I saw my primary care doctor, who sent me out for neurological and cardiovascular tests. When they came back negative, he asked about my hydration habits and told me I was most likely dehydrated. While I wasn't experiencing dehydration symptoms on a day-to-day basis, he said, whenever I put myself into a stressed state, that stress combined with a lack of proper hydration and electrolyte intake was causing me to faint. If I didn't do something to address it, I could end up fainting and hitting my head, which, aside from the embarrassment, could cause a concussion or far worse.

After that appointment, I started feeling light-headed everywhere I went. It was definitely psychosomatic, but I began carrying a water bottle with me at all times. It was the first time in my life I actually realized that if I didn't pay attention to proper hydration, bad things could and would happen.

In the years since my series of fainting spells, I've only had one other occurrence. I was in South America shooting a commercial for the reality television show *The Mole*, which meant I was outside of my regular routine of toting stainless steel water bottles and electrolyte mix with me everywhere I went. I was standing in front of a green screen when I started to get that familiar feeling of being light-headed. I didn't have a full fainting attack, thankfully, but we had to stop filming so I could lie down. The incident was my body telling me, once again, *If you're not going to work for me, I'm not going to work for you.*

Today, before I do anything that could be remotely stressful, I drink lots of water with electrolytes—I don't care what I'm doing or where I am. I also now carry water with me whenever I travel, give presentations, or appear on TV, no matter what kind of inconvenience it might be. I've learned that you can't take chances when it comes to proper hydration, which I now know means not just slugging water but also making sure you're drinking at the right times, balancing your water intake with electrolytes (if needed), and avoiding highly caffeinated or sugary beverages that can thwart your hydration attempts.

After I started working at the Amen Clinics, I had the opportunity to see what dehydration actually does to the brain. We were reviewing brain scans of professional body builders who had basically stopped drinking in order to cut water weight before a big competition. They'd given up water for only a handful of hours, but that was all it took to reduce their cerebral circulation significantly, which was surprising given how fit and healthy they were. Their scans were eye-opening, showing how immediately and profoundly dehydration can affect the brain.

This Is Your Brain on Dehydration

Three-fourths of all Americans are chronically dehydrated at any time.[1] This means the majority of us are walking around in a continual state of dehydration—not that you forgot to drink a glass of water one afternoon, but chronic dehydration.

Add this statistic to the fact our brain is made up of approximately 75 percent water—and needs to stay at 75 percent water to function optimally. Even losing 1 percent body weight to water loss can slash cognitive performance, interfering with memory, mood, mental energy, and focus.[2]

If you lose 2 percent of your body weight due to dehydration—still considered "mild"—your brain will downshift into dull. You will have slower reaction times, short-term memory problems, mental fatigue, confusion, anxiety, and impaired mood.[3] Mild dehydration can also interfere with motor coordination, increasing your likelihood of an accident—a frightening fact if you're driving, walking, or have a job in manufacturing or transportation that demands continual attention.[4] Making matters worse, studies show once you're dehydrated, its effects on mood persist after you fully rehydrate.[5]

Dehydration also shrinks the brain's precious gray matter.[6] Not drinking enough also drains cognitive performance, forcing your brain to do more work to process the same amount of information. Perhaps not surprisingly, well-hydrated people do better on cognitive tests, with improved memory, motor skills, mental energy, alertness, and concentration.[7]

From the body's perspective, dehydration is bad news for every ailment, including weight gain. Studies show people who meet or exceed daily water recommendations feel fuller and burn more calories than those who subsist in a state of mild dehydration. This metabolism uptick is no joke, either: research shows we can increase energy expenditure by as much as 30 percent simply by consuming sixteen ounces of water.[8] I notice, too, whenever I don't drink enough water, I'm a lot hungrier!

What happens if you let yourself get really dehydrated? Severe dehydration can cause extreme confusion and lethargy, and put you in a stupor, as your blood pressure drops precipitously. You can develop a fever, difficulty breathing, chest pain, and even seizures. Severe dehydration can also be fatal.

Many people assume that if they're not in a hot, humid climate, they don't need to worry about fluid intake since they're not losing water. But even if you remain in a climate-controlled room and don't move a muscle, you'll still expend water by being alive. In fact, the average person can lose more than a cup of water a day just through breathing alone! We also excrete about six cups of water daily through urine and bowel movements and eliminate about two cups through our perspiration.[9]

Taking prescription or over-the-counter drugs like antihistamines, laxatives, antacids, blood-pressure meds, and water pills can also put you at a greater risk of dehydration. If you're over age sixty, it's even easier to get dehydrated, since you have a lower thirst perception and your kidneys are less efficient at removing waste.[10]

Katy's Story

How Hydration "Changed My Life"

Katy reached out to me shortly after her son was diagnosed with postconcussion syndrome. He was only sixteen years old at the time, but after playing three years as the wide receiver–cornerback of his high school football team, he had developed severe brain fog, chronic headaches, and fatigue, and had trouble focusing on his schoolwork. After twelve months of rest, his symptoms weren't improving. While helping her son make changes to his diet and other habits—and get baseline tests like lab work, cognitive assessments, and brain imaging to assess his neurological status—I also started working on a brain-health program tailored specifically for Katy.

She told me she felt like something wasn't right with her neurologically. She had anxiety, her brain was racing, she was often dizzy, she wasn't seeing clearly, and she was forgetful. She told me she also had trouble retaining information and focusing on her computer screen. At fifty-three, she wondered if these symptoms were just part of turning fifty.

Typical of many in Los Angeles, Katy has a high-stress job, working as the CEO of a successful entertainment company. With so much going on in her life, she told me she didn't really think about drinking water or other fluids unless she was thirsty. We did the math on her daily hydration habits and realized that she was consuming no more than thirty-two ounces of water on average.

Instead of water, Katy was drinking several cups of coffee per day, along with diet cranberry juice and vitamin-infused waters. This meant she was consuming lots of artificial sweeteners, which are known to cause neurological problems. All the while, Katy actually thought these beverage choices were healthy, not realizing these juices and vitamin-infused waters are loaded with sugars, artificial sweeteners, and natural flavorings, which are not optimal for brain health.

At my recommendation, the first thing Katy did was get rid of all the drinks with artificial sweeteners at her office and home. At the same time, she invested in three 32-ounce stainless steel flasks and a juicer. She began filling the flasks with filtered water and carrying them with her wherever she went. If she craved a little zing, she'd squeeze some lemon or mint into her water, or add a splash of pomegranate juice. Also at my request, she began keeping a hydration journal, chronicling how much she drank, of which types of fluids, and when.

Instead of beginning her day with a cup of coffee, she started to hydrate herself with a big glass of freshly pressed green juice. She still had one cup of coffee in the morning, but later in the day, if she craved something warm and soothing, she'd have a cup of green or herbal tea.

In two months' time, Katy had turned around every neurological symptom she felt. Gone were the dizziness, fogginess, anxiety, memory loss, and difficulties concentrating. Instead, she felt like she was thinking more clearly, was sharper in conversation, could remember little details, and was overall happier, calmer, and more relaxed. She also told me she felt like her skin was glowing and looked younger. All of this because she finally stopped living in a state of chronic dehydration!

After eight months of daily green juicing, Katy was hooked, even launching an Instagram account dedicated to sharing the empowering effects she felt from the juice, using the hashtag #teamgreen. Her posts now attract hundreds of followers.

Today, every cognitive concern Katy had when we first met has gone away, which was through a combination of changes in hydration habits and optimizing deficiencies noted in her lab work, which helped us to tailor the rest of her brain health protocol. She continues to track her daily fluid intake in a journal, which she says helps her assess her hydration levels and keeps her accountable. She and her son have influenced the entire household, including her husband, to drink more so that everyone in the house is now hydrated and healthy. Katy says drinking more water—and getting rid of the fluids that weren't helping her brain—has changed her life and the lives of her family.

KRISTEN'S TIP: Don't try to hydrate with diet soda or other artificially sweetened beverages. While they may look like nothing but water with flavoring added, the chemicals and artificial sweeteners can interfere with cognitive function, thwarting your hydration goals.

This Is How Much You Need to Drink

There is no federal guideline for water consumption, in part because hydration needs are largely individual. The best general recommendation may be from the Institute of Medicine (IOM), which suggests men consume 3.7 liters (125 ounces) of water daily, while women are encouraged to consume about 2.7 liters (90 ounces) every day.[11]

The IOM's recommendation is a great goal, especially since most of us fall woefully short of that guideline. Some may also exceed the IOM's recommendation. I do, since I'm six feet tall and I exercise daily, which means I need to drink more for my gender. In addition, those who live in hot climates should also drink more, since you can lose up to sixteen ounces of sweat—the equivalent of a medium coffee—in just one hour if you're doing work, exercising, or playing outside in hot weather, according to sources.

If you exercise regularly, your hydration needs also increase. Being at high altitude can also deplete fluid levels since there's less oxygen, forcing you to breathe faster and lose water through exhalation. In addition, pregnant women, children, and older people also have specific hydration needs.

What you eat can influence your hydration too. On average, 20 percent of our fluid intake comes from food, but you may get less or more depending on your diet. Vegetables and fruits are highest in water content and can also hydrate better than water alone. That's because they contain natural chemicals like lutein and zeaxanthin that aid in hydration, along with sugar, electrolytes, and mineral salts for optimal fluid levels.

No matter what you eat, prioritize drinking water—and a ton of it. Even if you eat an inordinate number of vegetables, fruits, and servings of soup, foods can account for only a maximum of up to 40 percent of your total water intake.[12] Remember my story about graduate school? I was certainly eating a ton of produce, but I wasn't properly hydrated by any means.

How can you tell if you're drinking enough water? The easiest way is to take a look at the color of your urine. While that may sound unappealing, nothing works better for assessing hydration than a quick peek in the toilet. You may think of pee as universally yellow, but urine actually has a broad spectrum of hues, all of which can tell us a lot about our hydration and overall health.

For example, if your pee is a very light straw color or even clear, you're optimally hydrated. If your pee is darker than a light honey, you're mildly dehydrated and should drink water as soon as possible. Don't wait until your urine turns amber or even orange, which can indicate moderate to severe dehydration.

One note to keep in mind is that if you take any supplement with B vitamins like a multivitamin or a B-complex, the nutrients can cause your urine to turn a bright yellow. This doesn't mean you're dehydrated, but that your body is just excreting excess B vitamins.

Hydrate the Delicious Way

The following thirty foods are some of the most hydrating ones you can eat, with each containing at least 85 percent water per 100 grams, according to USDA calculations.

- **Cucumber**
- **Lettuce**
- **Grapefruit**
- **Celery**
- **Tomatoes**
- **Zucchini**
- **Watermelon**
- **Strawberries**
- **Cranberries**
- **Plain yogurt**
- **Spinach**
- **Cantaloupe**
- **Honeydew melon**
- **Kale**
- **Broccoli**
- **Peaches**
- **Carrots**
- **Oranges**
- **Pineapple**
- **Blueberries**
- **Bok choy**
- **Eggplant**
- **Apples**
- **Cabbage**
- **Raspberries**
- **Apricots**
- **Butter lettuce**
- **Broth-based soups**
- **Cauliflower**
- **Bell pepper**

The Electrolyte Equation and the Truth About Sports Drinks

Hydration isn't just about fluids. We also need electrolytes to stay hydrated. Electrolytes are minerals found in some beverages and food that help balance water levels throughout the body and brain while transporting nutrients into cells and exporting waste out. Our primary electrolytes include sodium, potassium, magnesium, chloride, calcium, and phosphate.

When we lose water through exercise, sweating, diarrhea, vomiting, or even a high fever, it can cause electrolyte imbalances. Certain medications, including some antibiotics and hydrocortisone drugs, and medical issues like a thyroid problem or eating disorder can also interfere with electrolytes.

The result of an electrolyte imbalance isn't pretty. Trust me: you don't want to faint whenever you're stressed out. You can also develop an irregular heartbeat, confusion, weakness, and excessive fatigue. Prolonged electrolyte imbalances can also cause damage to your nervous system and brain function and health. I'm so thankful I caught my problem early—even if it required a couple of fainting spells to do so!

Drinking enough water and eating a balanced diet is your first defense against electrolyte imbalance. If you work out every day like I do, consider an electrolyte mix without sugar or artificial sweeteners that you can stir into water. Natural drinks like coconut water or green juice (see pages 137–38 for more information on these two), which contain trace minerals and less sugar than commercial sports drinks, are better choices for replenishing electrolytes. A twenty-ounce bottle of a popular sports drink, for example, may contain up to 34 grams of sugar, 70 percent of the recommended daily value for a 2,000-calorie diet. This amount of sugar, unless you've worked out intensely or for hours on end, can impair cognitive health while contributing to increased hunger, weight gain, and other side effects.

The Dangers of Drinking Too Much

Three-quarters of Americans don't drink enough water, leaving them in a state of chronic dehydration. But sometimes, when people try to counter the problem, they end up drinking too much, resulting in water intoxication. Basically, this means you've had so much water that it dilutes your electrolyte levels, causing your cells to swell, and increasing pressure in your skull. This leads to headaches, confusion, and irritability, and sometimes also to nausea and vomiting.

When your blood loses too much sodium from overhydration, you can develop a dangerous condition known as hyponatremia. Eventually, hyponatremia can cause the brain to swell so much it can lead to seizures, coma, and even death.

Most of us don't have to worry about hyponatremia. But if you're an athlete, suffer from diarrhea, or have a medical condition like hypothyroidism, heart disease, or adrenal insufficiency, it pays to be cautious. Monitor your water intake, consume fluids with electrolytes after diarrhea or during prolonged exercise, and talk with your doctor about how to ensure your sodium levels stay on an even keel.

Why Water Quality Matters for a Better Brain

Our brain is 75 percent water—not 75 percent soda, juice, milk, coffee, iced tea, wine, beer, or diet soda. Water is the first and best choice for optimal brain hydration. And since our brain has no way of storing water, you need to continually drink it to rehydrate the brain.

What kind of water you drink, though, matters. Let's start with tap water. Much of the United States' public water supply contains contaminants that can interfere with basic health, including cognitive function.[13] Tap water has been found to contain toxic contaminants like lead, arsenic, fertilizers, pesticides, mercury, prescription drugs, and even harmful radioactive substances like uranium.[14] According to a recent study from the Natural Resources Defense Council, the public water supplies' top contaminants included disinfectants, lead, copper, and chloroform exposure at levels exceeding EPA regulations.[15]

Our tap water is also rich in chlorine, which utilities add in small amounts to kill bacteria and other germs. If you're drinking, cooking, and washing with tap water every day, your exposure can add up, impairing your central nervous system[16] and increasing your risk of cancer, kidney disorders, and skin irritation. Tap water also includes fluoride, which can help protect us from tooth decay but can also impede brain development and function.

Unfortunately, though, bottled water isn't much better. The same cancer-causing chemicals, toxins, and prescription drugs in tap water are also found in bottled water, oftentimes in higher quantities.[17] That's because safety regulations for bottled water are less stringent than public drinking water standards. While the EPA monitors and tests tap water frequently, bottled water is overseen by the FDA, which doesn't require lab testing or even violation reporting. So while we can get a good idea of what's in our tap water by looking up public testing information online, we have no idea what's in bottled water unless the manufacturer conducts and discloses independent testing data.

There's also the problem with the plastic bottles themselves. Plastic bottles can leach plastic particles and harmful estrogenic chemicals like bisphenol-A (BPA) in water, even if the bottles aren't exposed to heat or direct sunlight.[18] These chemicals, in turn, can damage brain cells, interfere with memory, and lead to mood disorders.[19] BPA-free bottles are no better and can still leak estrogenic chemicals into water, according to research.[20] Compounding the problem are growing environmental concerns about the nondegradable waste and other ecohazards created by the plastic bottles used to fuel America's obsession with portable water.

For all these reasons, I drink only filtered water out of glass bottles when I'm at home or stainless steel containers when I'm on the go. Purifiers or filtration systems are highly effective at removing many common contaminants, studies show, as long as you remember to change the filter or do any necessary maintenance on a regular basis.[21]

To find the right one for you, I suggest you start by finding out exactly what's in your tap water by going online or calling your water utility company for a copy of its annual Consumer Confidence Report.[22] You can then look for a filter or purifier that targets the specific harmful compounds in your tap water. For help with choosing a filter, visit the website for NSF International (https://nsf.org), an independent company that tests and analyzes consumer products like water filters. Whatever you buy, be sure it includes a seal of safety and efficacy from NSF, Underwater Laboratories, and/or the Water Quality Association.

I personally purchase water that has been filtered with a nano water purifier, which is pricey. But it's able to strain tiny nanoparticles that most home systems can't catch. The purifier also produces hyperoxygenated water, which contains more oxygen than regular water.

Choose Still, Not Sparkling, for a Better Brain

Sparkling water may be hugely popular right now, but I recommend avoiding it if you can. Since it's carbonated, sparkling water is more acidic, defeating the point of drinking alkalinized water. Carbonated beverages may also cause heartburn, gas, and bloating.

The Best Type of Water for Your Brain

People who know me also know I'm obsessed with the type of water I drink. And the best kind, in my opinion, is pure, filtered water, free of contaminants and slightly more alkaline than acidic. This means it has a higher pH level than tap water. Foods and beverages that are more alkaline may help neutralize acid in our bloodstream. While the evidence is inconclusive, many holistic health practitioners believe that the less acid we have in our body, the lower our disease risk. Studies also show animals who drink alkalinized water live longer.[23]

If you're not a fan of plain water, a quick fix is to squeeze lemon into your water. This helps infuse your water with vitamin C and phytonutrients, turning regular H_2O into a living water that can better fuel your body and brain. I always ask for lemon to squeeze into my water whenever I'm out at a restaurant or anywhere else I'm unsure of the water quality. It's a small addition, but the fruit helps deliver a nutrient boost to the water. If you're not a fan of lemons, you can add other fruits or vegetables such as orange slices, watermelon, raspberries, cucumber, or even a sprig of mint, to help enhance flavor and infuse it with the essence of living nutrients.

Three Beverages to Help You Win the Brain Game

Just because water is your body's preferred method of hydration doesn't mean it's the only thing you can drink. While most beverages contain sugar, artificial sweeteners, and/or other additives, there are a handful of drinks that happen to be extremely beneficial to your brain. Here are my three favorites.

1. **COCONUT WATER**: Think of coconut water as nature's original sports drink. The beverage, found naturally inside the coconut fruit, contains electrolytes without the synthetic sugar, artificial colors, and other additives found in commercial sports drinks.

   Coconut water also contains antioxidants like vitamin C that can help fight oxidative stress.[24] The drink can also lower blood sugar,[25] blood pressure,[26] and unhealthy cholesterol and triglycerides,[27] according to research. Personally, I think coconut water tastes so refreshing, giving you a flavor boost without any sugar or artificial sweeteners.

2. **TEA**: I drink tea all the time. Not only do I love the taste, I also find the act of making and consuming tea to be extremely relaxing—probably why it's been a ritual in parts of Asia for centuries. My favorite teas are organic green tea, peppermint, and decaffeinated cinnamon spice black tea.

   Tea also has amazing benefits for your brain. Drink green, black, and oolong tea, and you may be able to thwart cognitive decline by as much as 50 percent, according to recent research.[28] Other studies have shown that green tea can lower anxiety, boost memory, hone attention, and improve overall brain function and connectivity.[29] Drinking just one-half cup of green tea daily may lower the risk of dementia and depression while slashing the body's production of the stress hormone cortisol.[30] In fact, people who regularly drink green tea can lower their depression by as much as 21 percent, according to studies[31]—researchers say that's the stress-busting equivalent of doing 2.5 hours of exercise per week.

You can credit tea's incredible cognitive advantages in part to epigallocatechin gallate (EGCG), an antioxidant found primarily in green tea but also in black, white, and oolong. EGCG helps protect cells from oxidative stress while fighting inflammation[32] and has been shown to produce brain waves associated with relaxation and alertness.[33] For these reasons, tea, especially green tea, has been shown to help prevent a variety of conditions, including cancer, heart disease, diabetes, obesity, and neurodegenerative disorders like Alzheimer's.[34]

Green, black, white, and oolong tea also contain L-theanine, an amino acid that helps relax the central nervous system. And while they don't include anywhere near the amount of caffeine found in coffee, these teas have a small amount of caffeine, which helps to increase alertness and improve our mood.[35]

Herbal teas also play a protective role in the brain, capable of fighting off neurodegenerative disorders like Alzheimer's, according to research. For optimal health benefits, avoid adding milk, sugar, or artificial sweeteners to tea.

3. **GREEN JUICES:** I love green juice and drink one every day, no matter what, even if I have to send my fiancé, Mark, on a search to find one whenever we're traveling. Green juice is made by pressing whole, fresh green vegetables through a juicer. It's a a rich source of vitamins, minerals, antioxidants, enzymes, and phytochemicals, making it a nutrient dense beverage choice.[36] Green juice is also packed with

chlorophyll, the pigment that gives plants their green color, which helps detoxify and oxygenate the blood and lowers inflammation.

Your body also better absorbs the micronutrients in green juice than it does from the same green vegetables before they go into a juicer. That's because pressing veggies into juice breaks down cell walls and starches, allowing nutrients to be more readily absorbed. What's more, green juice doesn't contain fiber, which can bind to micronutrients and cause them to pass through our digestive tract without absorption.

Green juice is not a substitute for eating green vegetables. Instead, think of green juice as a good option when you want something flavorful, hydrating, and super healthy.

I drink a minimum of sixteen ounces of green juice every day, which I make at home in my juicer. You can use any green veggies you like, including kale, celery, spinach, swiss chard, arugula, broccoli, wheatgrass, parsley, cucumber, and cabbage. While my juice is primarily vegetables, I will add a serving of fruit, which may include blueberries, raspberries, strawberries, mango, pineapple, peaches, pears, or apples. A few more juicing tips:

- Make sure your juice contains more green veggies than fruit, the latter of which has more sugar and calories.
- Drink your juice within thirty minutes of pressing to prevent nutrient loss through oxygen exposure.
- Wash all produce before juicing and use only organic fruits and veggies to ensure green juice is free of pesticides and other toxins.
- Rotate the type of fruits and veggies you use for an optimal variety of micronutrients.
- Invest in a juicer. While you can use a blender, it won't remove fiber or pulp, so your produce will just be pureed, not juiced.

- When buying green juice at a café, make sure they're pressing the juice fresh for each order and aren't including added sugars, sweeteners, or other fillers.

I use a Breville juicer, but you can use any model that works for you. Here are two of my favorite green-juice recipes you can try at home.

MORNING HYDRATION BRAIN BOOST: I drink this one before I work out every morning. To make at home, juice 4–5 stalks celery, ½–1 whole, peeled cucumber, ½ cup Italian parsley, ½ cup baby spinach, and 2–3 stalks red kale or pacific kale. If you want the drink to taste a little sweet, add ½ to 1 whole green apple. Adding ½ cup cilantro can also help give the drink more detoxifying nutrients. Enjoy!

AFTERNOON KICKSTARTER: This juice is an abbreviated version of my Morning Boost for those afternoons when I'm rushing around and need a quick, easy energy kick. To make at home, juice 6–7 stalks celery, 2–3 stalks red kale or pacific kale, and 1 whole Bosc pear. Enjoy!

The Truth About Coffee

The reason we feel more awake and alert after a cup of coffee is because it contains caffeine. Caffeine raises blood pressure and heart rate, triggers the release of stress hormones, activates the adrenal system, and stimulates our central nervous system. That's a lot for the body and brain, which is why caffeine increases internal stress and anxiety.

While I appreciate the antioxidant benefits that can come from coffee, if you drink coffee throughout the day, you're constantly stimulating the body's stress response and central nervous system. Your body and brain are always on high alert, which, over time, can lead to reduced cerebral circulation, gray matter shrinkage, less neurogenesis, and diminished executive function and memory.

Coffee disrupts one of the things the brain needs the most: sleep. Even if you restrict your coffee consumption to morning only, it can still interfere with your sleep cycle, according to sources.[37] This is just one reason why studies show people who consume caffeine feel more fatigued than those who don't drink it at all.

Roasting coffee can also result in the production of acrylamide, a chemical that in high concentrations can impact the nervous system.[38] Coffee also blocks the body's absorption of magnesium, a mineral critical to brain health in which many people are deficient.

If a client tells me they need a cup of coffee to wake up, I suggest we look at other areas of the body that might need support, like diet, hydration, exercise, and nutrient deficiencies. Most people tell me they eventually perform better and feel more alert when they start their mornings with a large glass of water instead of a big mug of coffee.

My Favorite Coffee Alternative: Spicy Focus Tea

I drink this health elixir whenever I want a flavorful hot beverage that can also stimulate my mind. It's rich in brain-healthy minerals and tastes like a chai tea.

TO MAKE: Boil 1 cup hot water. Remove from heat and add 2 tbs unsweetened organic coconut powder, 1 dropper liquid trace minerals, 1 tsp chaga mushroom extract powder, and 1 tsp Sunfoods Golden Milk Superfood Blend (optional). In lieu of the Superfood Blend, spice to taste with turmeric, cinnamon, ginger, cardamom, and/or black pepper. If you like it sweet, you can add a few drops liquid stevia.

The NFL Story

HOW GIVING UP COFFEE BROUGHT MORE MENTAL CLARITY TO THIS NFL HALL OF FAME PLAYER

Ed White, a former Hall of Fame offensive guard for the San Diego Chargers and Minnesota Vikings was sixty-two when he came to see me as part of his brain-health evaluation. After spending seventeen years in the NFL, during which time he won numerous division championships, made four trips to the Pro Bowl, and played in the Super Bowl, Ed was concerned about his mental health and worried he wasn't as sharp as he used to be.

Like many players I worked with, Ed paid little to no attention to the fluids he was consuming and what effect they might be having on his brain. After we sat down together, I quickly realized this former football star was consuming too little water—and far too much caffeine and sugary beverages. Ed told me he would often drink up to four cups of coffee a day, which he admitted didn't make him feel so healthy. He also told me he drank water only when he was thirsty and liked to enjoy soda whenever he could.

After I thoroughly explained to Ed how coffee was hijacking his central nervous system and interfering with his cognitive health, he was immediately open to changing his hydration habits for the sake of his brain. He decided to wean himself down from four cups of coffee per day to only one, which today he brews to be half regular coffee and half decaf.

In place of coffee, Ed started consuming more

tea. Because he liked the taste and ritual of coffee, I thought Ed would enjoy drinking tea, since both are similar in the sense that they offer a variety of rich flavors, blends, and preparation techniques, all of which can go into producing the perfect cup. Ed loved tea and started drinking green tea and herbal blends like orange tea. He even bought a teapot so he could steep loose-leaf tea in order to brew a fresher, more flavorful cup.

At the same time, Ed stopped drinking soda and invested in a forty-ounce stainless steel flask, which he loaded up with ice and filtered water. He began taking the flask with him wherever he went, even if it was just to another room in his house, and finishing several full flasks throughout the day into the evening. He began leaving the flask by his bed, drinking from it at night if he woke up.

These habits, combined with other tweaks we made to his diet, exercise, and supplemental regimen, helped to change Ed's mental performance in a matter of weeks. While giving up coffee was difficult at first, he soon realized it was more of a habit than a need, and he began to notice he had more mental clarity and felt less jittery without it. In addition, he preferred the even-keeled energy he got from drinking green tea.

More so, consuming more water has helped Ed feel mentally sharper, more focused, and healthier overall. Because he's hydrated, he also has more stamina for his daily hour-long walks up into the mountains where he lives. He says water keeps his appetite in check, helps him focus on his artwork, and gives him the energy boost he needs to be 100 percent around his grandchildren.

KRISTEN'S TIP: Many people have a difficult time giving up coffee because the ritual of a morning cup or stopping to get a latte on the way to work is so ingrained in their schedule. Ease the transition like Ed did by substituting a tea ritual for your old coffee routine.

7

THE STRESS OFFENSIVE

Here's the truth of it: we all have stress. More than 80 percent of all Americans deal with stress on a daily basis,[1] making the United States one of the most stressed-out countries in the world.[2]

But regular stress is different from chronic stress. Your brain on regular stress is like a car with a flat tire, whereas your brain on chronic stress is like driving a car cross-country when the alternator is shot, the oil is low, and the timing belt is frayed or cracked: it's probably just a matter of time before the engine blows. From a cognitive perspective, stress kills nerve cells, shrinks gray matter, ruins your ability to think clearly, and significantly increases your risk of age-related decline, dementia, and Alzheimer's.

I can admittedly be a pretty anxious person—and have been since I was a child. At the age of five, I developed a nervous disorder known as trichotillomania, and started manically pulling out my hair to try and relieve whatever stressful emotions I was feeling. The condition upset my mother, especially when people stared at her small blond-haired daughter with unusual bald spots.

While I eventually grew out of that particular anxiety-related behavior, my stress levels increased in graduate school, with a full class load and time in the lab—which interfered with my sleep schedule. I tried to counter the pressure with exercise, which certainly offered some immediate relief, but it didn't prevent me from developing a form of restless-leg syndrome, kicking and jerking my legs at night, and grinding my teeth while sleeping. I started experimenting with ways to reduce my stress levels with weekly acupuncture and a consistent meditation practice, and felt that with those additional tools I was in a good place.

When I started working at the Amen Clinics, I was excited to see the results of my first brain scan as I was proud of the fact I had been consistently practicing great brain habits. While I certainly knew I had periods of stress-related anxiety, I was eating healthy, exercising daily, and taking high-quality nutritional supplements to support my brain. Even so, to my surprise, my brain showed considerable electrical activity in areas associated with stress. To turn down the dial on my stress, I started doing yoga several times per week and tailoring a supplement regimen geared to reduce anxiety.

This is all to say: you can follow the Better Brain Diet to a T, exercise daily, take every supplement I recommend, hydrate continually, and do sudoku puzzles until your eyes cross, but if you're under too much stress, you won't be able to effectively boost your brain function and health.

As you know from chapter one, stress impairs cerebral circulation, prevents neurogenesis (the growth of new brain cells), and overactivates neurons that, over time, create new neural pathways that can interfere with cognitive function. Too much stress also shrinks the brain's hippocampus, limiting memory formation and recall, while increasing the strength of connections to the amygdala—the brain area associated with processing emotion, causing us to experience more fear and anxiety. These changes can be lasting, too, limiting our ability to think clearly, solve problems, make smart decisions, stay focused, and be happy and healthy over time.

Stress isn't only what we feel on a mental and an emotional level. We can also incur stress from chronic ailments like diabetes, arthritis, and high blood pressure, along with unhealthy habits like eating too much sugar, getting too little sleep, not exercising enough, and failing to hydrate. What's more, every one of us faces environmental stress from pollution and radioactive materials in the atmosphere, in addition to toxins in our food and water.

You can't control all sources of stress. But you can do everything you can to mitigate that stress by getting enough sleep and practicing specific therapies like yoga, meditation, and deep breathing.

The Sleep Offensive: How Much You Really Need for a Better Brain

Ninety-nine percent of the time, I can fall asleep anywhere, anytime, often in a matter of seconds. I can put my head down on a desk and doze, I can fall asleep on planes and in the passenger seat of a car, and I can peacefully drift off to sleep on the couch. I even fell asleep in the middle of the *Terminator* movie! (As enjoyable as it was, it just goes to show when my brain wants to shut down, it does.)

The other 1 percent of the time, though, I can't fall asleep at all. Not one wink of sleep. Instead, I lie awake the entire night, not even nodding off for a brief moment. This is a complete departure from my norm, but there's a pattern: this happens whenever I have a stressful event—like a big presentation—the following day.

If I can't fall asleep, I don't beat myself up about it—this only increases the likelihood I won't ever sleep. Instead, I try to focus on what I can do to relax my body and allow it to get some rest. I stay in bed, with my eyes closed, and I don't turn on the lights, read, watch TV, or look at my phone. If I can, I try to meditate or concentrate on letting my body relax while clearing my mind. Through it all, I tell myself that I'm still doing something beneficial for my body by allowing it to "sleep," even if my brain isn't shifting into the deeper states of relaxation to get the rest it needs.

Despite this practice, I'm still mentally shot the day after a sleepless night. I can't think clearly, I struggle to come up with ideas, and I tend to get irritated more often. The tone of my voice even changes. I'm usually a happy-go-lucky person, but when I don't sleep, it's much more difficult to have a resilient, optimistic view of the world.

While your brain never shuts down, sleep is the only time it gets to recharge and repair itself. This is when the flow of cerebrospinal fluid increases, washing away harmful toxins and waste products that may otherwise build up. An excess accumulation of this waste—which includes beta-amyloid, the protein found aggregated in those with Alzheimer's—is one reason why research links sleep disorders to the disease.[3]

The brain also consolidates short-term memories and the new knowledge we learn during the day into long-term memories while we sleep. That's why you may notice that, if you don't get enough ZZZs, you can't remember something someone just told you the day before. For this reason, people perform significantly better on memory tests after a single night's sleep or even a short nap, studies show.[4] Students who sleep more also get higher grades, according to research.[5]

Similarly, sleep is vital to executive function—or the ability to plan, make good decisions, stay organized, and remain focused. Too little sleep can also make you think you're right when you're wrong and prevent you from making good decisions. We also need adequate sleep to conduct high-level thinking and have those "a-ha" moments that help turn random thoughts into exciting new ventures and multi-million-dollar businesses.[6]

If you don't get enough sleep, certain areas of the brain can become overactive, creating new neural pathways that impair thinking, concentration, and cognitive efficiency. Getting less than seven hours' sleep also increases stress and anxiety by as much as 30 percent[7] and boosts the risk of depression ten times more than people who are adequately rested.[8] Some sleep researchers even say if we all got just sixty to ninety minutes more snooze time every night, the world would be a happier place.[9]

Inadequate sleep doesn't just affect the brain, obviously. Too little snooze time can cause weight gain and high blood pressure, in addition to increasing the risk of everything from wrinkles and the common cold to fatal car crashes, heart disease, cancer, diabetes, and premature death.

So how much sleep do you really need? Leading national health institutions recommend at least seven hours per night—some people may need up to nine hours, depending on activity levels, lifestyle habits, and health needs.

We've all heard those myths about the successful CEO or president who slept just four hours per night. But the four-hour sleep myth is just that: a myth. There is one caveat: a miniscule percentage of the world's population is known to be short sleepers, meaning they can sleep fewer than six hours per night and still feel refreshed in the morning, thanks to a rare genetic mutation.[10] Short sleepers, however, make up only 1 to 3 percent of the total population.

While short sleepers are rare, the chronically sleep-deprived among us are not. Forty percent of Americans get less than seven hours of sleep on average.[11] Most people also overestimate how much sleep they get, and those who are the most sleep deprived are also the most likely to inflate their snooze times.[12] That's because not getting enough sleep impairs your judgment across the board, including your ability to self-assess.[13]

Six Ways to Get at Least Seven Hours' Sleep Per Night

1. **GET ON A SLEEP SCHEDULE.** One of the best ways to increase both your sleep quantity and quality is to go to bed and get up at the same time every day, including on the weekends. Doing this resets your body's internal clock so you tire at the same time every night, cuing your body to prep for sleep.

2. **TURN DOWN THE TEMPERATURE.** Your body lowers its core temperature to help initiate sleep, so turning down the thermostat or opening a window can help stimulate the process. The ideal room temp for sleep is between sixty and sixty-seven degrees Fahrenheit.[14]

3. **IGNORE YOUR PHONE, TABLET, TELEVISION, AND LAPTOP.** You've heard this advice before, but that doesn't stop people from staring at their phones right up until or even after they turn out the lights. But the blue light emitted by smartphones, laptops, computers, tablets, and televisions can stimulate the brain, preventing you from falling asleep. Instead, say good night to your technology at least ninety minutes before you want to

go to bed. If you can't make this commitment, invest in blue-blocking glasses or use an app that filters blue wavelength.

4. ADOPT A SLEEP RITUAL. In my house, we dim the lights around 8:00 every night, helping cue the body to start producing melatonin. Around 9:30 P.M., I'll take our rescue dog, Oscar, on his last walk, brush my teeth, wash my face, and crawl into bed with a good book, which I'll read for a few minutes before shutting off the light.

5. STOP EATING THREE HOURS BEFORE BED. Eating dinner or snacking too close to bedtime can cause indigestion and prevent you from falling asleep and staying there.

6. TAKE A BATH WITH SLEEP-ENHANCING INGREDIENTS. When I think I might have trouble falling asleep, I draw a warm bath with Epsom salt. When dissolved, the salt releases magnesium, which penetrates the skin to calm and relax muscles and nerves. Adding a few drops of the essential oil lavender to a bath can also help promote sleep, according to research.[15]

Sleep Apnea: The Secret Brain Killer

Sleep apnea causes people to have disrupted breathing when they sleep, preventing the brain from getting enough oxygen. The disorder affects approximately 25 million in the United States—one out of every twelve Americans—and significantly raises the risk of cognitive impairment, dementia, and Alzheimer's. Symptoms include loud snoring, waking up feeling like you're gasping for air, a sore or dry throat in the morning, and daytime drowsiness. Talk with your doctor if you suspect you might have sleep apnea. Getting treated for the condition can increase your cognitive function overnight, literally, and save your brain from decline and disease.

Kristy's Story

THE POWER OF YOUR PRACTICE TO RESHAPE YOUR BRAIN

You may remember Kristy from chapter one. She appeared so placid and calm on the outside, but her brain images revealed a different story. Despite her tranquil temperament, Kristy's scan showed excessive activity in areas associated with anxiety. The stress wasn't just affecting her brain—she'd also been recently diagnosed with ulcerative colitis, which meant her mind-body connection was playing out profoundly in her gut.

There's one big part of Kristy's story we didn't talk about in chapter one—her poor sleep habits. Not only did she have trouble getting to sleep, she also had trouble staying asleep. After sleeping for six or less hours, she'd get out of bed feeling fatigued and low-energy, even though her brain felt overstimulated.

I asked Kristy to commit to a consistent sleep schedule, so she decided to try going to sleep at 10 P.M. and waking up at 7 A.M. Instead of staying up until she felt tired, Kristy started preparing for sleep every night around 9, dimming the lights, powering down her phone, using essential oils like lavender to relax her mind, and taking magnesium and GABA thirty minutes before she climbed into bed. She also started taking probiotics to help support her gut. These rituals worked, helping improve her sleep quantity and quality, and she started to average seven hours of restorative sleep per night.

Once we had her sleep back on track, I suggested we take the next step: managing her daily stress through meditation and breath work. Whenever she got overwhelmed, Kristy began breathing deeply into her belly, concentrating on a positive affirmation like "I am calm." The practice—along with adding meditation and yoga—also made her feel like she was in control of her stress rather than the other way around.

Kristy continues practicing yoga, meditation, and deep breathing to this day. She says she's also been more creative, coming up with solutions to problems rather than staying stuck in them. She says it's like someone took the top off a teapot and let the steam out—all her negativity, restlessness, and anxiety have been released.

KRISTEN'S TIP: Taking her yoga and meditation practice outside whenever she could helped Kristy lower her stress even more. Studies show spending just twenty minutes in nature, whether you're practicing meditation, yoga, or deep breathing, can help calm the mind significantly.[16]

Three Ways to Train Your Brain for a Calmer, Smarter Mind

To better control your stress over time, you need to find an activity that can lead to functional and structural changes in your brain. According to research, meditation, yoga, and deep breathing all rewire the brain's neural networks, making them powerful long-term tools. Even doing one of these activities a few times a week can have a significant impact on your brain.

1. Meditation

Some who haven't tried meditation are skeptical of the practice at first. I was more curious initially. Like many other logical, linear thinkers, I assumed meditation meant sitting in a lotus pose with your eyes closed, trying to quiet all thoughts in your mind. However, after years of developing my own meditation practice and having studied the research on how impactful meditation is on brain function, I have an entirely new appreciation of this technique in optimizing brain function.

The definition of meditation is broad, and there are many schools of thought on how to practice. Personally, I define meditation as taking a moment to sit still, do an internal inventory, and practice getting into a space of being more mindful, focusing on the present moment at hand. For many, this means finding a quiet space to be alone and concentrating on your breath and body. Thoughts might enter your head, but instead of mulling them over, you acknowledge them and return to your breath. This practice is generally known as mindful meditation.

How does mindful meditation change the brain? If you practice daily, you can learn to shift brain wave states, from our alert, attentive, waking state (beta brain waves) to a quieter, more relaxed resting state (alpha brain waves). If you practice for weeks or months, meditation increases the brain's gray matter, including in the hippocampus—the brain's memory center—helping you better regulate emotions, according to research.[17] Meditation also reduces the size of the amygdala, the brain's fear center, decreasing how much stress and anxiety we can feel.[18] Interestingly, the practice also quells activity in the areas of the brain responsible for self-referential thoughts, which is the tendency to relate all occurrences, big or small, to oneself.[19] Self-referential thoughts increase worry and anxiety, but meditation creates new connections in the brain that limit this kind of mind wandering. It also helps significantly reduce the stress hormone cortisol and markers of inflammation.

According to research, people who meditate for just two months have less stress and anxiety a full three years after they stop practicing.[20] Meditation also makes us smarter, increasing our mental clarity and ability to focus and make good decisions. Finally, meditation can help heal many physical ailments, including high blood pressure and chronic pain, while reducing your risk of Alzheimer's and other neurodegenerative disorders.

HOW TO DO IT: Find five to twenty minutes when you can retreat to a quiet place away from social, audible, or visible distractions, and sit with eyes closed. Focus on your breath, inhaling and exhaling slowly. If your thoughts distract you, don't get upset. Simply acknowledge them and return to your breath, releasing negative thoughts or emotions out with every exhale.

Don't worry if meditation feels uncomfortable at first. Be patient and commit to trying it several times before you make up your mind about the practice. You can also download a meditation app like Headspace to help guide you. Many meditation newbies have had great success with apps, which are often free or low-cost.

TRY MEDITATION IF . . . the idea of sitting still for a few minutes sounds like an extended vacation, you value your alone time, or you feel like you can't get out of your own head.

The NFL Story

HOW THIS 65-YEAR-OLD HALL OF FAME STAR USED MEDITATION TO SACK STRESS

Clinton Jones was a former Michigan State Hall of Fame running back, Heisman Trophy nominee, and a first-round draft pick of the Minnesota Vikings, where he spent a majority of his seven-year career in the NFL. When we met, he was long past his days of competing in the Super Bowl. He was in his mid-sixties, working as a chiropractor, and under a different kind of stress than he felt during his days as a competitive athlete. At times the stress of his job and life made him feel anxious and disorganized, akin to a rat on a sinking ship, without a glimpse of driftwood in sight. He also wasn't sleeping well, averaging four to six hours per night, during which time he would often wake up.

After he described his sleep habits, we suspected sleep apnea and recommended Clinton get tested. Like many football players we had treated at the clinic, his diagnosis was confirmed and he was prescribed a continuous positive airway pressure (CPAP) device to help treat it.

Sleeping longer and more deeply certainly helped Clinton, but the next step was finding a practice to help him manage his stress. He had been interested in meditation for years, so I suggested he double down on his current practice—and add breath work—all with the goal of boosting his cognitive health.

Clinton followed my advice more than I could have imagined, meditating daily, sometimes twice a day. He discovered a meditation technique that worked for him, which included chanting a sound out loud while focusing on a visual object like a mandala, often used in Buddhist and Hindu cultures.

The effects of Clinton's practice were inspirational. Here was this sixty-five-year-old former pro athlete, meditating for up to two hours daily. That commitment immediately changed his stress levels, his sleep patterns and—most important—his perception of life events. Negative ideas that before had bothered him for days or even weeks soon became passing thoughts, as he was able to become more mindful both inside and outside his practice.

Today, Clinton meditates for a minimum of an hour every morning and night. He tells me he wouldn't be alive without it—it helps him to stay balanced physically, mentally, and emotionally. He credits meditation with transforming every cell inside him and giving him more courage and strength than all his years on the football field.

KRISTEN'S TIP: There are lots of stereotypes and misconceptions about meditation. The practice isn't just for New Age types. Many top CEOs, politicians, celebrities, and athletes, including pro football players, practice the therapy. Like Clinton, you should find a form of meditation that works for you and be willing to experiment and have an open mind.

2. Yoga

After seeing my first brain scan and being surprised by how my stress levels were impacting my brain, I started taking vinyasa flow yoga classes in the morning before work. After committing to a sixty- to ninety-minute practice several times a week, I found myself learning about what it meant to be truly present in the moment, focused on my body and not on the thousands of thoughts flowing through my mind. Instead of worrying about work, my commute, my father, my love life, or all the other things we can get hung up on, I concentrated on learning the poses and executing them to the best of my ability. I also loved it as a form of gentle stretching, which my body desperately needed to counter the effects of all of the running that I do.

Similar to meditation, yoga increases gray matter[21] and decreases the size of the brain's amygdala, not just when you practice but for days afterward.[22] It stimulates production of GABA, which helps calm anxiety—and this isn't just simply because you're doing something relaxing. According to one study, an hour-long yoga class boosted GABA by 27 percent, while reading quietly for the same amount of time had no effect on the neurotransmitter.[23] Yoga also lowers cortisol and increases serotonin, dopamine, and other feel-good hormones.[24] It calms activity in the brain's frontal lobe,[25] considered to be our cognitive control panel, and increases production of brain-derived neurotrophic factor (BDNF), the protein linked to neurogenesis, the brain's ability to grow more brain cells in adulthood.

HOW TO DO IT: The simplest way to start is to find a certified yoga instructor or studio and try a few classes. Many gyms and health centers also offer yoga as part of a basic membership. There are dozens of different types of yoga, and classes can vary in length and difficulty, so be sure to find the right class for your schedule and comfort level. If you don't want to join a studio or take a group class, there are hundreds of apps and online courses that can help guide you through the practice wherever you choose to do it.

TRY YOGA IF . . . You feel more relaxed when you're moving your body, the idea of focusing on adopting specific positions seems calming, or you like the community aspect group yoga can provide.

3. Deep Breathing

What I like about deep breathing is that you can do it anytime, anywhere, and see immediate results. It's been shown to lower cortisol levels within seconds,[26] bringing down blood pressure and heart rate at the same time.[27] This means it can calm someone down from a crisis situation almost immediately. Deep breathing, also called diaphragmic breathing and belly breathing, has been used to treat phobias, motion sickness, posttraumatic stress disorder, and other stress-induced emotional disorders.[28]

I use deep breathing whenever I'm anxious about something or need to calm my mind in an instant. For me, deep breathing flips a switch, short-circuiting the electrical activity in my brain.

HOW TO DO IT: Place one hand over your belly and the other hand on your heart. Inhale deeply through your nose for a count of eight, bringing the air into your belly, then hold for a count of four. Exhale slowly through your mouth for a count of eight. If you find a count of eight is too taxing, lower both your inhalation and exhalation duration to a count of six. Repeat five to ten times. You can also close your eyes, which some people find enhances their relaxation.

Why to Invest in Massage: Self-Care for Your Body and Brain

I'm always hunched over my microscope or computer—a posture many office workers can relate to. While there are countless detriments to sitting too long, especially with poor posture, a big one for your brain is that the habit creates tension in the neck and shoulders. This constricts blood vessels, reducing the circulation of blood and oxygen to the brain, leading to headaches, brain fog, and other cognitive problems.

Massage can relieve the tension of sitting too long while lowering cortisol and soothing your sympathetic nervous system.[29] A good massage session can also reduce blood pressure, heart rate, and cortisol, increase production of endorphins and serotonin, and help you sleep better at night, regardless of when you receive it.

If you think you can't afford massage, I've got good news: my favorite place, around the corner from my home, costs $25 per session—that's less money than most people spend on coffee drinks every week. Ask friends, colleagues, or your family doctor for a recommendation or head online to community review sites like Yelp to find an affordable masseuse near you. Keep in mind getting a massage doesn't need to be a weekly occurrence—you could even schedule one once a month, if you know you have a stressful event coming up.

TRY DEEP BREATHING IF . . . You suffer anxiety around events like interviews, air travel, or public speaking, you want an instant stress-reliever, or you like the idea of fighting anxiety no matter where you are (for example, driving in your car, standing in line at the store, fighting with a loved one, etc.).

12 Other Amazing Ways to Reduce Stress and Relax

Here are some of my favorite ways to turn down the dial in my head.

1. **GO FOR A JOG.** Running is my favorite form of moving meditation. It helps quiet my mind and organize my thoughts.

2. **SPEND TIME WITH ANIMALS.** Forget diamonds—dogs are a girl's best friend. If you don't have a furry friend at home, consider volunteering at an animal shelter for some added animal time.

3. **CALL YOUR BEST FRIEND.** Whenever I chat on the phone with my closest girlfriend, even if only for a few minutes, I always laugh. She brings a different, refreshing perspective to my problems.

4. **GO TO THE BEACH (OR LAKE, RIVER, FOREST, FIELD, PARK, OR MOUNTAIN).** Nature has a science-backed, quantifiable effect on the brain, reducing stress and improving positivity and mood.

5. **TAKE A GABA SUPPLEMENT.** I use the supplemental form of this calming neurotransmitter whenever I feel overwhelmed.

6. **TURN UP THE HEAT.** If you've been in a steam room or jacuzzi, you know they're incredibly relaxing for your muscles and mind. Bubble baths and heating pads are great alternatives.

7. **READ AN UPLIFTING BOOK.** Whenever I read a motivational book, it quiets my thoughts, puts things into perspective, and inspires me to be a better person. On my bedside table now: *The Little Book of Inner Peace: The Essential Life and Teachings* by the Dalai Lama, and *Be More Dog: Life Lessons from Our Canine Friends* by Alison Davies.

8. **USE AROMATHERAPY.** Apply the essential oil lavender to your wrists or put it in a diffuser to use this science-backed scent to calm your nerves.

9. **DOODLE OR DRAW.** Being creative can help channel stress into something beautiful and positive. I used to love drawing pictures of horses to soothe my sympathetic nervous system. Now, I doodle brains, flowers, and geometric shapes.

10. **REVISIT HAPPY MEMORIES.** Curate a little digital—or physical!—album of photos/notes that bring you joy, and look at it every time you're doubting yourself.

11. **LISTEN TO A RELAXING SOUND.** For me, it's the Om chant or crashing ocean waves. For you, it might be classical piano, choral music, rain, the human heartbeat (you can download an app for the last two), or any other sound that soothes you.

12. **SEE THE LIGHT.** Getting outside light helps calm and energize the brain, which is why I prefer to be in rooms with lots of windows. If I have to be in an office without windows, I'll install lights with full-spectrum light bulbs, which emit a luminescence similar to natural sunlight.

8

THINKING YOUR WAY TO A BETTER BRAIN

My father was the paragon of positivity, even though some might say he had no reason to be. He'd seen a remarkable amount of death and trauma in his lifetime, having served two tours as a combat helicopter pilot in Vietnam and after spending twenty-five years in the fire service. But these experiences never diminished his optimism. Until his death, he remained sunny and hopeful, always looking on the bright side, even when things seemed dire. He was positive about other people, too, never saying an unkind word about anyone and trying to see the best in everyone.

Growing up, I not only learned the powerful benefits of optimism from my father, but that positivity could make you happier and healthier, too. I didn't realize at the time that optimism could also improve brain health in incredible ways.

If you wonder what outlook has to do with cognitive function,

I get it. It took me some time to realize just how much our thoughts influence our brain, not just our mood and mental health but also how well the brain actually works.

Quite simply, every thought we have influences cognitive function. Think negative thoughts, and you'll activate hormones, neurotransmitters, and structures that lower clarity, creativity, focus, problem solving, decision-making, and the overall ability to think and process. Think positive thoughts, and you'll activate different hormones, neurotransmitters, and brain regions that lead to higher cognitive function, better decision-making, and a happier disposition.

Proof of positivity can be found in research on people who live the longest. The world's longest-living humans hail from all over the globe. Dan Buettner's five "Blue Zones" with the highest per capita centenarians are in Icaria, Greece; Nicoya, Costa Rica; Sardinia, Italy; Okinawa, Japan; and Loma Linda, California. These centenarians have different genes, eat divergent foods (although most diets are semivegetarian, plant rich, and whole foods based), and subscribe to an array of habits and religious beliefs. But the common aspect they all have is that they're optimists, spending most of their (long) lives thinking positive thoughts.

On average, optimists live an average of 11 to 15 percent longer lives, according to research.[1] Scientists believe that positive people are more adept at regulating emotions and behaviors to make better decisions and handle stress.

But can you wake up one day and decide to be an optimist? And can you really change your thoughts to change your brain?

In a nutshell, yes. I know because I've seen people do it and the benefits their brain has reaped as a result. No one can avoid all negative thoughts or self-doubt, and the practice of changing one's thoughts, like most things worthwhile, takes dedication and patience. But the potential payoff for your cognitive power and performance is immense.

How the Real-Life Jerry Maguire Turned His Life Around with Positive Thinking

Many people know Leigh Steinberg, if not by name, by association. One of the most successful sports agents ever, Leigh was the real-life inspiration for the title character in the movie *Jerry Maguire.* Today, he's the CEO of Steinberg Sports and Entertainment, which has represented more than three hundred professional athletes worldwide, including Troy Aikman, Warren Moon, and 2020 Super Bowl champion Patrick Mahomes.

When I started working with Leigh in 2010, he was recovering from alcoholism and in financial ruin, having lost most of his net worth. His father, whom he had been close to, had recently passed away, his marriage was fraying, and his two sons had both been diagnosed with an eye disease that can cause near blindness. He was at rock bottom and had trouble believing a light out of the darkness even existed.

When we met, Leigh wanted to work on his cognitive health, but in order to do that, I knew we had to address what was taking up most of his mental energy. He was overwhelmed with negativity and didn't understand what effect it was having on his cognitive function.

I taught Leigh that negative thoughts damage brain cells and create new neural pathways that impair the ability to think and see things clearly and accurately. This was impactful for him. He had once been one of the world's savviest sports agents. How could he achieve that again if he couldn't even think clearly?

During our sessions, I focused on helping Leigh become more mindful. I asked him to put away his problems for a few minutes and focus on what was happening in the moment instead. Outside our sessions, I asked him to practice mindfulness whenever he felt like a drink or a dessert, since he had started substituting sugar for his alcohol addiction. He also began journaling his thoughts.

I asked Leigh to visualize his goals and picture himself achieving them. This sparked some optimism inside him, as he began to work toward his ambitions. He started to believe there might be a light flickering somewhere in his dark world. Positive thoughts and images helped rewire the neural pathways he needed to have hope.

By changing his thinking, Leigh transformed his life. When we first met, he couldn't think clearly. In a matter of months, he felt he could perceive the world as it was again. He rediscovered gratitude and suddenly saw his troubles for what they were: insignificant by comparison. Instead of sinking into self-despair, Leigh was bolstered by the things he did have: he was healthy, he had his family, and he had tools to make his life what he wanted it to be.

With optimism and hard work, Leigh became a successful agent again, rebuilding his company to secure more than $3 billion for his athletes. He has also directed $750 million to charities—the effect of his newfound gratitude. But for Leigh, his most meaningful comeback was becoming a good parent again and maintaining his sobriety.

Things aren't always going to be perfect for Leigh, but he understands that's OK—that's life. He now knows how to light candles and find the solutions rather than remaining in darkness.

KRISTEN'S TIP: Practicing gratitude is an easy and effective way to overcome negative thoughts and what they can do to our cognitive power. You don't have to be rich, famous, or successful to be grateful. If you're alive and relatively healthy, that's every reason to count your blessings.

How Our Thoughts Change Our Brain

The number of thoughts humans on average think every day is a subject of great speculation. Some sources claim we think sixty thousand independent thoughts daily,[2] while others like spiritual practitioner Dr. Deepak Chopra assert the number is closer to eighty thousand. However many tens of thousands of daily thoughts we have, up to 90 percent are repetitive and up to 80 percent are negative.[3]

Repetitive thoughts occur when you think the same thing over and over again, whether you're reliving the past, anticipating the future, or considering what's happening around you. Repetitive thoughts can be positive—for example, when you savor a fond memory, anticipate an event you're looking forward to, or prepare for something in the future.[4]

But repetitive thoughts become troublesome for our brain when we ruminate, or dwell on the same negative thoughts over and over again. We do this when we continually worry about something in the future or past that makes us feel sad, guilty, insecure, angry, or dark. These kinds of thoughts, in turn, can lead to depression, anxiety, and other problems in the brain and body.[5]

While many repetitive thoughts are negative, we can also have negative thoughts that aren't repetitive. In the brain, negative thoughts increase stress, boosting cortisol and internal inflammation. Over time, this uptick in stress hormones and inflammation produced can damage the hippocampus,[6] impacting our ability to think, recall information, solve problems, be creative, and mentally perform at our best.[7]

Negativity also lowers activity in both the cerebellum, which helps control thinking and motor skills, and the temporal lobes, which can lead to memory issues, impulse-control problems, and mood disorders.[8] Worse still, negative thinking overactivates the amygdala, the brain's fear center, creating dark emotions and causing us to store present experiences as bad memories.[9]

Perhaps unsurprisingly, negative thoughts have been shown to increase the risk of depression, anxiety, bipolar disorder, and nearly every other mood dysfunction. They also drive up the likelihood of Alzheimer's and other forms of dementia.[10]

Every thought we think has the ability to rewire the brain's neuronal connections and synaptic strength, eventually creating new neural pathways. Similar to repetitive thoughts, the more negative thoughts we have, the more negative neural pathways we create. Negative thoughts can even change genes, shortening telomeres—the protective caps at the ends of chromosomes—which causes cells to age more quickly.[11]

Our thoughts shape our emotions—and our emotions drive our decisions. If your thoughts are negative, your emotions will be, too, causing you to base your decisions on a negative outlook. Consequently, negative thoughts lead people to make poor decisions that can create or worsen bad situations.

Repetitive and negative thoughts also increase the risk of heart disease, diabetes, disability, cancer, and other chronic ailments.[12]

The good news: positive thoughts have the exact opposite effect. Thinking good thoughts cuts stress, increases cognitive function, and improves mood—in other words, optimism make you smarter, happier, and healthier. The effect is so powerful that positivity has been associated with everything from increased pain tolerance[13] to helping people fight the common cold.[14]

In the brain, positive thoughts lower cortisol and inflammation, and increase feel-good neurotransmitters like serotonin and dopamine that stimulate feelings of calm, focus, and relaxation. Positivity also activates both the prefrontal cortex, which helps regulate thoughts and emotions, and the hippocampus, which increases cognition and learning.[15] Positivity also lengthens telomeres, slowing down aging, and helps us make better decisions and find solutions to problems.

Optimism also profoundly affects the body, decreasing the risk of chronic illness while increasing mortality and quality of life. One study even found people with a family history of heart disease who also have a positive outlook are one-third less likely to suffer a heart attack or other cardiovascular event than those who have a more negative outlook.[16]

If you assume you think more positive thoughts than negative ones, you're not alone: most people do—and most are wrong. A study conducted on business school students found that while the students expected their thoughts to be 60 to 75 percent positive, in reality, negative thoughts accounted for 60 to 70 percent of their total thoughts.[17]

The Placebo Effect: The Incredible Power of Your Positive Thoughts

Perhaps nothing better demonstrates the power of positivity than the placebo effect. The placebo effect describes the benefits people receive from an inactive pill or treatment because they believe that treatment will produce positive consequences. In other words, if you believe something will work, the power of your positive thought will make it work.

The placebo effect is no joke. Over the past several decades, more and more research has shown that placebos can work as well as prescription drugs and certain medical interventions in treating conditions like chronic pain, depression, sleep disorders, menopause, and symptoms of Parkinson's disease. The placebo effect has even helped people defeat heart disease, cancer, and arthritis. In his book *You Are the Placebo,* Dr. Joe Dispenza details how he was able to avoid surgery and possible paralysis after breaking six vertebra simply by using his mind to heal his body.[18]

While there are many stories showing how placebos promote health, there is an equally comprehensive amount of research on the subject. For example, research with brain imaging has found placebos activate the same areas in the brain that prescription drugs do in patients suffering from chronic pain and Alzheimer's.[19] Other research shows prescription antidepressants are no more effective than placebos in treating mood disorders.[20] Similarly, scientists sent shock waves through the pharmaceutical industry when they released findings showing that placebo sugar pills reduced chronic pain as effectively as some of the most powerful painkillers on the market.[21]

How can placebos produce these incredible effects? Patient expectation is the number one reason placebos are effective. When we believe and expect a pill or treatment to work, it creates a powerful connection between the body and mind. Our mind gets excited thinking we will heal, which causes the body to release endorphins and other neurotransmitters that stimulate the healing process. At the same time, your cortisol levels drop in expectation that your pain or suffering will end soon, which has a therapeutic effect on the body, decreasing pain while lifting mood.[22]

But the placebo effect works only if you believe the therapy or rituals around the therapy will work. In other words, you need an optimistic attitude that a pill or treatment will accomplish the results your practitioner says it will. For this reason, studies show optimists respond better to both placebos and biologically active drugs or treatments.[23]

The NFL Story

HOW MERRIL HOGE USED THE POWER OF HIS THOUGHTS TO CRUSH CANCER

Former NFL running back Merril Hoge wasn't my client, but we've been friends for several years. I've included him because I love he how he used the power of positivity to turn a dire diagnosis into a renewed passion for life.

After leaving the NFL in the mid-1990s, Merril was diagnosed with stage II non-Hodgkin's lymphoma at age thirty-eight. His physicians weren't sure if chemotherapy would cure his cancer or if he would ever fully recover from the disease. He even wondered how he was going to find the strength to get through cancer.

Merril's doctors recommended an intensive course of chemotherapy for six months to one year. The treatment would be brutal, they said, without any guarantee it would work. The idea of dying began to overwhelm him, and he allowed himself to become consumed by thoughts of disease and death.

Merril decided to tell his kids immediately about his diagnosis and prepare them for the changes to come. Upon learning what was happening with her father, his nine-year-old daughter, Kori, threw her arms around his neck, peered up into his face, and said, *Dad, find a way.*

In that moment, Merril's energy changed. Dying was no longer an option.

"Find a way" had been Merril's favorite slogan as a player and father. After he heard his own no-excuses saying from his daughter, he knew he had lost his drive and courage. How would he survive cancer if he couldn't find a way? He realized he had to change his thoughts and see cancer like he would any other obstacle on a football field—something he simply had to overcome.

Merril took action. When he found himself thinking he would die from cancer, he flipped the thought into a positive *I will* destroy *non-Hodgkin's lymphoma*. He didn't know how he was going to do it, but it didn't matter: instead of dying, he became consumed with the idea of beating cancer. He started visualizing his organs as healthy, clean, and cancer-free.

Merril started thinking whom he could find who had been through the disease—he wanted support and proof of life on the other side. With the help of his doctors, he connected with several patients who had survived lymphoma, and he listened to their stories of how they had conquered cancer. He felt heartened that everyone he met had made it. Sure, some were still bald and they had crawled to cross the end zone to get their touchdown, but they had all won the game.

With his reformed attitude, Merril was in remission six months later—a half year before doctors expected—and was back to work as a commentator at ESPN soon after. His recovery was determined, daring, and incredible to himself, as well as to his doctors.

Seventeen years later, Merril is still cancer free. And he still reframes negative thoughts into the idea, *Find a way*. The words continue to inspire him, helping him feel powerful and in control of his circumstances, no matter what the world throws him.

Today, Merril writes down all his goals and pins them

to a corkboard, reviewing them in the morning and night and thinking of steps to take to achieve them. The board has become part of his process, combining optimism, visualization, accountability, and action in one place. Every day he tells himself to do something that will take him closer to accomplishing his goals.

Optimism has had a ripple effect across Merril's life. Cancer was not his high point, but the disease helped instill in him the power of positivity, which has made him happier, able to think more clearly, be more creative, and focus on what he wants to do and who he wants to be.

KRISTEN'S TIP: Unfortunate circumstances may, understandably, make us want to give up, but they offer us an opportunity to transform ourselves.

Eight Steps to Finally Break Free of Negative Thinking

Negative thoughts are normal. We all have them, even people like me who are perpetual optimists. I can get caught in the mind spin of negativity when I overanalyze things, which I'm prone to do. It's part of my scientific nature, but it becomes problematic when overanalysis leads to paralysis and I can't accomplish anything because I think it's not perfect. I overprepare for events that need just five minutes' prep or put things off to the last minute because I'm anxious everything needs to be executed perfectly.

I sometimes tell myself I can't take on a new project at work, for example, because I think I won't be able to do a flawless job. Or I can't serve a meal for friends unless it looks and tastes exactly like the cookbook depicts. These thoughts paralyze me and prevent me from pursuing new goals and dreams.

When paralyzed by a case of *I can't*s, I recognize I'm in a negative thought pattern and make a conscious effort to reframe the situation. Instead of thinking, *I can't take on a new project because it won't be perfect,* I tell myself, *I can do this project and am excited to give it my best. Nothing is perfect, but I'm confident and eager to confront the unknown.*

Instead of thinking *I can't make dinner for twenty people,* I tell myself, *So what if the meal doesn't look or taste like the cookbook? What is the worst that will happen? I am happy to be with friends and family and be able to share a homemade meal with them. I can always run down to my local natural-food store and get something catered if I need a backup plan.*

While you can't calm all negative thoughts (nor should you try—it's an unrealistic feat), you have a choice about how you approach life. Here's how to quell negative thoughts in eight steps and work to be smarter, happier, and healthier.

1. Journal Your Mental Chatter

You can't change thoughts if you don't know what your thoughts are. Keeping a record of the thoughts that occur in your head when you're washing the dishes, driving your car, running errands, commuting to work, walking your dog, or otherwise running on autopilot can help you identify negative patterns.

To keep a thought journal, write down as many thoughts, images, or words that come into your mind during the day as you can. Since tens of thousands of thoughts enter our head on a daily basis, focus on those that change your emotions or mood, even if slightly—that is, if a thought causes feelings of worry, sadness, insecurity, inadequacy, anxiety, or irritability. Also include automatic negative thoughts, or when you tell yourself you can't do something, you blame yourself, or you use words like "always," "never," or "I should" in a negative context.

After you've journaled for several days to one week, read through your journal to identify patterns and which negative thoughts keep resurfacing. Are you blaming yourself? Are you predicting what will happen in the future based on one event? Are you giving ultimatums, like you'll never find a new job, you'll always be alone, and so on? Look for similarities in circumstances when these thoughts occurred: Do they happen at work or around job-related issues? Do they occur when you're alone or with friends, family, your spouse, or colleagues?

When you identify a common negative thought, make a list of reasons it might be true. If you think you'll always be alone, for example, write down evidence for the idea. Then, list all the reasons you may not always be alone: you could meet someone, another person could ask you out, you could develop a new friendship or live with family to take the place of an intimate relationship in the interim. When we do this, we usually end up seeing that our negative thought is based in emotional overreaction, not reality.

2. Reframe Your Negative Thoughts

After you've identified a few common negative thoughts, you can begin to stop yourself when you think them and reframe them into positive sentiments.

How you reframe your thoughts depends on the thought itself. For example, if you think you're not good enough, look at the evidence for why you think this. If you're comparing yourself to others, remind yourself we all have unique backgrounds and experiences, and whomever you're comparing yourself to is likely comparing him- or herself to others, too. (Also remember that things are never as good as they look on social media—more about this in Step Five.) Reframe *I'm not good enough* to *I'm OK just as I am. I have limitations like everyone else, but I also have dreams like everyone else, and I am excited to give my best effort to attain my dreams and be the best person I can be.*

Sometimes it helps to reframe a negative thought into something proactive or actionable. For example, if you feel guilty about something you're doing or not doing, like not losing weight, not exercising, or not being able to complete a project at work, instead of saying, *I should do* (or *not do*) *X, and I'm a failure because I can't,* reframe to, *I intend to do* (or *not do*) *X by taking these steps . . .*

3. But Don't Try to Control Your Thoughts

Negative thoughts are natural, and trying to control them or stop them altogether can harm you more than it can help. If you tell yourself, *No! I shouldn't think that!* every time you have a negative thought, it can increase worry and overall negativity. Instead, accept your negative thoughts without judgment and ask yourself why you're feeling negative about yourself or your circumstances. Go through the same steps you did when you journaled. What evidence do you have that the thought is true? What evidence do you have that it's not true? Finally, challenge the negative thought by reframing it into something positive.

4. Quiet Your Brain by Practicing Mindfulness

Learning to be mindful and turning your attention to what's happening in your body and brain in the present can prevent you from ruminating over the past or trying to predict the future. Mindfulness not only quiets our thoughts, it also trains us to become more aware of our emotions and mediate them more effectively in the future.

Whenever I'm overwhelmed or negative, I take myself out of my current situation and find a quiet space to meditate. I close my eyes and turn my focus inward, paying attention to the inhalation and exhalation of my breath and the corresponding physical sensations in my body. If a negative thought enters my head, I acknowledge it and recenter my focus back on what I'm doing now—sitting alone in a quiet space and redirecting my thoughts to my body.

When I don't feel like sitting, I take my mindfulness outside and run, walk, bike, or hike. Activities that include repetitive motions allow you to sync your breath with your steps and can help you better focus on what's happening in your body as you move through the moment.

5. Rethink How You Consume Social Media, Television, and News in General

Most of us are constantly tethered to outside information and what we see on social media and our phones, computers, tablets, and TVs. The problem with all this information is that much of it is negative and, in the instance of social media, can create a vicious cycle of comparison and pessimistic thoughts.

If you spend a lot of time on social media, it can be easy to start comparing yourself with what you see there. This can spark negative thoughts, that you're not successful, thin, athletic, adventurous, rich, or loved enough. The more time we spend on social media, studies show, the less happy and satisfied we feel with our own lives.[24] Remember too that social media presents only a curated view of someone's best life. We rarely see the hard times, insecurities, uncertainties, loneliness, and failures.

The more time people spend on social media, the less time they spend exercising, pursuing hobbies, hanging out with family and friends, and doing other activities that bring meaning and happiness to their lives. Social media consumption also increases social isolation and feelings of loneliness and depression.[25]

News has a similar effect. Most news is negative—there's a reason the saying *If it bleeds, it leads* exists among news editors. Watching TV news in particular can increase anxiety and sadness and cause people to catastrophize their own concerns, according to studies.[26]

Limit the time you spend on social media and watching TV, which has stronger emotional impact than reading news. Even on-air shows like dramas and reality TV are often negative or violent and can increase stress. If you want to unwind, do something that will truly relax you like spending time with friends, reading a good book, taking a bath, or exercising.

6. Remake Your Morning Routine into a Positive Start

If you start your day positive, you're more likely to stay that way. This is why some of the world's most successful people like company CEOs, presidents, and scientists have morning routines they'd never give up, making sure they exercise, spend time with family, meditate, or read before they start their busy day.

I start each day with a glass of pure water and a green juice I press myself before going for a run or doing some other type of exercise. When I finish, I meditate for five to twenty minutes and then grab some quick snuggle time with our dog, Oscar. This routine physically, mentally, emotionally, and spiritually fortifies me, preparing me to handle anything that comes my way.

7. Move Your Body

You know from chapter four that exercise reduces stress and improves self-esteem and confidence. It can also help you break the cycle of automatic negative thoughts, research shows.[27] For me, exercise is one of the most reliable sources of positivity. Working out almost always lift my spirits and makes me feel better about myself and life in general. When I'm not trying to be mindful, exercise also helps me work through conflicts and turn over negative thoughts. We're also more creative and better able to solve problems when we exercise, according to studies.[28]

8. Just Make the Easy Choice

Everyone has a choice. You can either be optimistic and look for solutions to your problems or you can stay negative and wallow in situations, allowing them to stagnate or worsen. To me, the choice is obvious. Negativity serves no purpose—it never makes a situation any better and usually only worsens whatever problems you do have.

Don't Be Alone with Negative Thoughts

A licensed psychologist or therapist can help anyone better deal with negative thinking, whether you have a mood disorder or just want to be the best person you can be. One of the most effective treatments for negative thoughts is cognitive behavioral therapy (CBT), which works to address underlying thought and behavior patterns that cause pessimism. CBT can help you understand where negative thoughts originate and how to reframe them into positive affirmations or proactive solutions. To find a therapist, ask for recommendations from family, friends, colleagues, your primary care doc, or others you trust. You can also go online to the American Board of Professional Psychology and search under "cognitive behavioral therapy" for a certified specialist in your area.[29]

9

THE BRAIN GAMES YOU REALLY NEED

You might have heard that puzzles like crosswords and sudoku stimulate the brain and help ward off cognitive decline. But brain games—or cognitive training, as we neuroscientists like to call it—do more than make you *slightly* sharper as you age. Cognitive training has the potential to target multiple aspects of your cognitive power and improve them in a matter of months, boosting memory, concentration, comprehension, problem solving, creativity, and even intelligence. If that weren't enough incentive, brain games can also help treat cognitive damage (which we discovered during our clinical trial with NFL players) and slow the brain's aging process.

I grew up in a household of brain games, even though I didn't know it at the time. My mother played lots of solitaire and gin rummy, and she made us the go-to house in the neighborhood for weekly bridge games. Our cabinet was also full of board games like Trivial Pursuit, checkers, and backgammon, along with word games like Mad Libs (my favorite) and three-dimensional combination puzzles like Rubik's cubes.

Our cognitive training didn't stop with traditional games, though. My mother was an avid artist and spent her free time drawing, painting, sculpting, and weaving, often inviting me to join her or encouraging me to take classes at the Art Institute of Chicago. She just loved creating, and whether that was through art or baking, I learned alongside her. I couldn't tell you the number of times I stood next to her in the kitchen, measuring ingredients, following family recipes, and watching her estimate cooking times and fine-tune cooking methods.

Through my father, I was introduced to a myriad of musical instruments, everything from the classical guitar and harmonica to the piano, flute, tambourine, banjo, and the child's classic, the recorder. I tried to learn from sheet music and the different ways to play each instrument, also tinkering with how to tune and clean each one.

This is all to say I was constantly surrounded by pursuits that challenged my cognitive capacity, introduced me to new skills, and made me think in different ways—and that is exactly what cognitive training is all about. It doesn't have to take the form of a traditional "game" to help stimulate the growth of new brain cells, strengthen neural pathways, and sharpen your mind in the moment and for years to come.

The research behind cognitive training is equally impressive. One recent study, for example, found that adults who play ten hours of video games total over a period of time lengthen their cognitive reserve—the brain's ability to function despite some damage—by as much as three years.[1] Other studies show playing brain games for just a handful of weeks can affect cognitive function up to a decade later.[2]

One of cognitive training's most intriguing benefits is its potential to increase IQ.[3] While the research is still evolving, some studies suggest the more brain games you play, the smarter you make your mind.

Brain games also spur neurogenesis, creating new brain cells as we age.[4] Doing activities that challenge the brain also increase neuronal connectivity, stimulating new pathways that help us think more efficiently and effectively.[5] And as you likely know, brain games can help stave off Alzheimer's and other forms of dementia.[6]

I saw firsthand what brain games can do for cognitive function while I was running a clinical trial with NFL players at the Amen Clinics. As part of the brain rehabilitation component of the trial, we gave the players a thirty-minute basic neurocognitive assessment that included twenty-nine brain-training games, which were individualized to the cognitive areas they needed to improve. The players could then play the games at home using a computer program, which we encouraged them to do on a daily basis.

The games appealed to their competitive nature, and many excelled at the training. Knowing we would assess their performance at the six-month mark, they were motivated and committed to playing at home and improving their skills. That dedication proved extremely effective, with a majority of the players improving their cognitive function and proficiency to some degree, and nearly half of the players demonstrating an increase of up to 50 percent or more.

Four Tips to Get the Most Out of Your Brain Games

Chances are you're already doing some cognitive training on a regular basis. But because the brain constantly needs to be challenged in new ways to stay sharp and healthy, you'll want to mix up your cognitive workouts. Here's how to master cognitive training to biohack your brain:

1. **EMBRACE THE NEW.** If you do the same thing every day, your brain gets bored. If you do crossword puzzles daily, for example, your brain grows accustomed to the challenge and will eventually no longer be stimulated to grow. Similarly, if you've played the violin for years, learning the viola won't be as beneficial to your brain as if you tried to pick up the trombone. Try new things to keep your mind challenged, young, and healthy.

2. **RETHINK IDLE TIME.** Waiting for a plane to take off, a train to show up, or a daily commute to be over can be mind-numbing. Turn it around by transforming boredom into brain power with cognitive training on the go. If you're at the airport, pick up a napkin and try writing with your nondominant hand. If you're waiting in your car to pick up a child or spouse, download and play a brain-training app (BrainHQ is my favorite). If you're driving, challenge yourself to remember as many items in a single category (e.g., types of dogs, flowers, famous artists, etc.) as possible in one minute.

3. **SWITCH IT UP.** Even if it's new to you, playing only computerized brain games or doing only crossword puzzles won't boost your cognitive function as much as engaging in multiple activities. Experts compare it to exercise: If you lift weights with only your arms, your legs won't get strong and you won't boost your cardiovascular system.

4. **BE CURIOUS.** Make a conscientious effort to tap your inner curiosity and learn more about our wonderful world. Educate for the sake of pure joy and the love of knowledge, as well as for your cognitive power and health.

Janet's Story

HOW BRAIN GAMES CREATED A BETTER PICTURE FOR A RESTLESS MIND

When I met first Janet, she was struggling with anxiety and insomnia and wanted to know if there was more she could do to calm what she described as her continually restless mind. The fifty-five-year-old also had a demanding job as the CEO of a successful electronic media company and was concerned with protecting her neurological health after her father passed away from the progressive neurodegenerative disease amyotrophic lateral sclerosis (ALS).

A scan of Janet's brain corroborated what she was feeling inside. Certain parts of her brain were overactive, with high beta-wave activity in areas responsible for organization, spatial orientation, cognitive processing, attention, and procedural memory (the ability to remember how to do things and complete tasks). Her results told me that she needed to train her brain to increase her attention, memory, and overall intelligence, which would help hone her cognitive processing. More important, she needed to do something stat to reduce her anxiety.

Janet already liked to do crossword puzzles and things by hand rather than typing or texting. I suggested she double down on crosswords and also start playing word-search games, which would help build her vocabulary and boost her intellect. I also encouraged her to find ways to use her hands to calm her mind, whether she wanted to draw, paint, knit, or write with her nondominant hand. And since she loved puzzles, I suggested jigsaw puzzles, which I know from firsthand experience can calm even the most anxious mind.

After our session together, Janet started doing crossword and word-search puzzles every day, saving time-consuming jigsaw puzzles for the weekend. That was when she learned her mind could really relax. While she had never considered herself a jigsaw puzzle enthusiast before, she fell in love with the game. She started with 100-piece puzzles before working her way up to 500- and 1,000-piece sets and eventually advanced to 3,500-piece puzzles.

Eventually, jigsaws became a form of active meditation, letting Janet focus her mind on the moment, giving her more peace and clarity. While she puzzled, she said she would suddenly come up with solutions to her problems without consciously thinking them over. For Janet, who was passionate about real estate, the puzzles were like renovating old houses, giving her the chance to play the game of what-should-go-where as she worked to envision a cohesive, more beautiful structure.

Within weeks of beginning her brain-game regimen, Janet told me she felt more stimulated and engaged and had improved her concentration, not to mention her vocabulary. More important, she was more relaxed than she'd been in months and was sleeping more consistently through the night. She was using the games to unwind and reduce stress while also improving her cognitive function.

Brain games are now an essential part of Janet's life. She says she looks forward to getting home from work to start a new jigsaw puzzle and can lose herself for up to eight hours at a time in a single game. She keeps a stack of crossword books and word-search games on her kitchen table for inspiration, doing them daily or whenever she flies or can't sleep at night. For Janet, the games have been transformative, not only for her brain but also for her quality of life.

KRISTEN'S TIP: Brain games don't just hone the mind—they can also provide a critical outlet for stress release. To get the biggest bang from your brain game, find a pursuit like drawing, puzzling, or playing music that you find as soothing as you do challenging.

10 Brain Games for a Sharper, Smarter, Healthier Mind

We all have certain areas of our cognitive capacity we want or need to work on, whether it's our mental clarity, attention span, memory, or general intelligence levels. For me, I'm always trying to improve my mental efficiency—I value a quick, nimble mind that's able to rapidly assimilate information. Given the amount of reading I do, I also try to train my brain to improve my reading retention and comprehension. Here, I've outlined the best brain games for ten different cognitive goals. Find yours and turn whatever cognitive deficiency or desire you might have into your greatest mental strength.

1. **IF YOU WANT TO BOOST INTELLIGENCE . . . READ FOR THIRTY MINUTES EVERY DAY.** We all have three types of intelligence: crystallized intelligence, or the accumulation of knowledge, facts, and skills; fluid intelligence, or how well we reason and solve problems regardless of what we know; and emotional intelligence, or how well we respond to individuals and in social situations. Reading, especially long-form narratives

like books, for at least a half hour every day is the best way to boost all three, experts say.[7] I'm sure you already read plenty of e-mails, texts, social media posts, and work memos, but getting engaged in an actual story for at least thirty minutes will increase activity across the brain, improving your overall neuronal connectivity and the integrity of your white matter tracts, according to studies.[8]

2. **IF YOU WANT TO IMPROVE YOUR MEMORY . . . MASTER A NEW WORD EVERY DAY.** When I was young, I used to wrench my parents' big dictionary out of our bookshelf, sit down, and scroll through it, looking for new words to learn. Today, I still do the same thing, although without the heavy lifting, using the Merriam-Webster Word of the Day app to learn a new word every day. Today's word, for example, is "parvenu," which is someone who recently rises to an unaccustomed position of wealth or power without the prestige to go with it. See? So much fun!

   Mastering new vocabulary boosts working memory, which is an asset of our short-term memory that's central to both our basic memory and overall intelligence.[9] Because working memory has a limited capacity, expanding it by learning new vocabulary helps us to communicate more efficiently and creates new ways to retain more information over time.[10]

3. **IF YOU HAVE ONLY FIVE MINUTES . . . PLAY A COMPUTERIZED BRAIN-TRAINING GAME.** What I love about computerized brain games like BrainHQ and Lumosity is that you can play them anywhere, anytime. I'm always pulling out my phone and playing a quick game on my app while waiting for a friend, for an exercise class to start, or for Mark to scan the menu when we're out for dinner!

My favorite computerized brain-training app is BrainHQ. Not only is it clever, easy to use, and entertaining, BrainHQ was also found to be the most effective for cognitive training by independent researchers who ranked the eighteen most popular computerized brain-training programs.[11] What's great about the app is that you can also choose which cognitive skill you want to hone, whether it's your memory, navigation, spatial orientation, cognitive speed, intelligence, attention, or focus.

4. **IF YOU'RE WORRIED ABOUT DEMENTIA . . . LEARN A NEW LANGUAGE**. You likely already know language is one of evolution's greatest gifts to the human brain. But fascinating studies now show the act of learning a new tongue can stave off dementia by years.[12] By comparing people who speak one language to those who are bilingual, researchers found that people who can speak more than one language develop dementia later in life on average than the monolinguals, even though the latter group tends to have more formal education.[13]

   Don't have time to learn a whole new language? That's OK. Even memorizing foreign words without committing to picking up the tongue can help prevent cognitive decline. My father was Swedish—both of his parents were born in Stockholm—so I enjoy memorizing new Swedish words and phrases to help keep my mind sharp.

5. **IF YOU WANT TO TRAIN YOUR BRAIN TO BETTER HANDLE STRESS . . . BECOME AN ARTIST**. Whether you like to paint, draw, sculpt, take photos, knit, weave, throw pottery, or do any other artistic pursuit, making art increases cognitive capacity in ways other brain games do not. Producing visual art, for example, has been shown to increase functional

connectivity across the different areas of the brain, making us more psychologically resilient to stress.[14] Artists also have more gray matter on both sides of their brain—not just the right side, which has been associated with creativity—increasing connectivity to help them better handle complexity and crisis.[15]

Doodling has similar cognitive benefits, especially if you flip whatever you're drawing upside down. It may sound odd, but the method helps better integrate the right- and left-hand sides of the brain, making you more mentally adept and nimble. Former NFL offensive guard Ed White (see page 142) loves to draw images upside down with his dominant hand, then re-create them right side up with his nondominant hand—a total creative challenge for the brain!

6. **IF YOU WANT TO FIGHT AGE-RELATED COGNITIVE DECLINE . . . VOLUNTEER.** Most people don't think of volunteerism as a brain game, but it is a brain enhancer. Studies show being philanthropic can help prevent and even reverse age-related volume loss in certain areas of the brain like the hippocampus.[16] My grandmother, who lived to the amazing age of ninety-five, volunteered regularly at a hospital for forty-five years, which I strongly feel was the reason she stayed so cognitively sharp and healthy later in life. Volunteering on a regular basis also lowers stress, feelings of depression, and anxiety,[17] while boosting overall well-being,[18] all of which have been shown to help counter age-related mental decline.

7. **IF YOU WANT TO GROW NEW BRAIN CELLS . . . TAP YOUR INNER KEATS.** Writing creatively, whether it's a story, a poem, a limerick, a love letter, a diary entry, or anything else expres-

sive, increases the size of the hippocampus by growing new brain cells. According to studies, this occurs because writing challenges the brain continually to come up with words and create new ideas.[19] Writing by hand also activates multiple parts of the brain, improving thinking, language, and idea generation.[20] Whenever I want to commit something to memory, I write it down, even if I'm in a lecture or meeting where a laptop would be easier.

8. **IF YOU WANT TO HONE YOUR FOCUS AND ATTENTION . . . DO A CROSSWORD, JIGSAW PUZZLE, OR SUDOKU.** These three games all require you to focus on words, puzzle pieces, or numbers in order to solve the problem, which, if you play often enough, will boost your attention span. In fact, studies show people who do crosswords and sudoku on a regular basis have similar cognitive capabilities to those ten years younger.[21] Unlike some cognitive-training exercises like computerized brain games, which usually have a specified time limit, you can easily spend hours immersed in a difficult puzzle or numbers game. My fiancé, Mark, just gave me a complicated jigsaw puzzle of the different dog breeds that I can't wait to spend the weekend completing!

9. **IF YOU WANT TO INCREASE YOUR MENTAL CLARITY . . . TAKE A DIFFERENT WAY TO WORK.** Every time you take a new route, even if it's just turning right at a light where you normally make a left, it challenges your brain, increasing your gray matter and ability to focus, think, remember, and learn, all of which will improve mental clarity. The strongest evidence of this comes from a study conducted over a decade ago on London taxi drivers, whose brains were compared to those similar in age, education, and intelligence who didn't drive taxis. Researchers found the taxi drivers had significantly

larger hippocampi because they were constantly taking new ways around the city of some twenty-five thousand streets.[22] The longer a taxi driver had been working, the bigger his hippocampus was, according to the study.

Another way taking the road less traveled increases clarity: you're more likely to be mindful when you do. Whenever you take a new route, it forces you to notice new surroundings, keeping you grounded in the moment and what you're doing at the present time.

I try to take new routes all the time, using the Waze driving app to discover different streets in my neighborhood—you'll be surprised at the options. By taking new routes, I've also discovered some incredible new restaurants, parks, places to walk our dog, Oscar, and other local gems, which has deepened my appreciation and enjoyment of where I live and work.

10. **IF YOU WANT A SIMPLE WAY TO CHALLENGE YOUR BRAIN EVERY DAY, NO MATTER WHAT YOU LIKE, WHERE YOU ARE, OR WHAT TOOLS YOU HAVE AT YOUR DISPOSAL . . . JUST TRY SOMETHING NEW.** Many brain games have the same objective, which is to learn something new. Even if I didn't detail one of the ways you like to learn, anything you can do to pursue a new skill or an area of knowledge, whether it's listening to a TED Talk, trying a new recipe, taking a golf lesson, or watching a video on a subject you know nothing about, will help stimulate your brain and increase your cognitive power and performance.

For me, I love to listen to podcasts on new neurology studies from the *Journal of the American Medical Association* and news segments that summarize the day's events from the *New York Times*' podcast, *The Daily*. Find your passion and pursue it for a healthier mind and a smarter, healthier brain.

The NFL Story

THE SECRET BRAIN GAME THAT'S REVOLUTIONIZING HOW ATHLETES PLAY AND THINK

Jon Vincent, a former long snapper at the University of Cincinnati, was first introduced to neurovisual training (NVT) during his first year as a Bearcat. He quickly became impressed with the training, so much so that he decided to change paths and earn a B.S. in neurobiology, eventually pursuing a career in the neurovisual industry after graduation. We met at a training camp for young hockey players in L.A., where Jon, along with physicians in neurology and ophthalmology, could teach the players about a type of brain training no high school coach was likely to show them.

What exactly is NVT? Neurovisual training uses simulators, computer screens, and virtual reality headsets to hone athletes' eye movements and overall optical skills. Computer programs and games challenge their ability to process and absorb complex movements while increasing their ability to register and react to split-second actions—all vital functions for high-performance athletes. An ocular workout like no other, the training also strengthens the eyes' motor muscles, helping to prevent strain, headaches, blurry vision, and double vision.

But NVT doesn't just sharpen your vision. The "neuro-" part of neurovisual training is just as important. NVT works the brain intensely, strengthening the critical pathways between the eyes and brain while improving how quickly we process visual information. Studies show NVT increases attention, working memory, and visual information and processing speed.[23] For all these reasons, rehab clinics around the world use NVT to help patients recover more quickly after traumatic brain injuries.

Today, NVT is used at universities and by professional sports teams around the country to increase their athletes' peripheral awareness, dynamic visual ability, depth perception, hand-eye coordination, decision-making, and focus. With NVT, coaches now realize that no matter how strong or quick players might be, they can go only as fast and as far as their brain can tell them to go. Since NVT was first implemented at UC ten years ago, the number of concussions has dropped by an incredible 80 percent, thanks in part to the training's ability to bolster situational awareness.

When Jon attended UC, he and his teammates were required to complete two hours of NVT every week for the six weeks of preseason. Once the season began, the players did thirty minutes every week simply for maintenance. In this sense, NVT was no different from strength training and conditioning, but instead of working the body, athletes would spend hours training their brains.

For Jon, NVT was invaluable to his success as a college athlete. As a long snapper, he was constantly trying to outrun 250-pound linebackers hell-bent on taking him down from any angle. Before he started NVT, he would regularly get blindsided on any given punt. But after the training, he began to see threats in his peripheral vision before they even appeared, processing them quickly and effectively to sidestep a game-ending tackle.

Today, Jon still does NVT regularly. As a result, he says, he thinks more clearly, makes decisions more efficiently, better comprehends fast-moving information, can focus more easily, and rarely gets bogged down by random thoughts.

KRISTEN'S TIP: The NVT systems used by universities, professional sports teams, and rehab clinics are pretty expensive, but since I was first introduced to the science, there's been an explosion of at-home versions, which are far more affordable and accessible. To find a personal NVT set, call an ophthalmologist in your area who sells the products or can recommend where to find one.

10

BIOHACKING YOUR BRAIN IN REAL TIME

You cannot change what you do not measure.

This was the maxim we lived by at the Amen Clinics. Think about it: if you don't know whether anything is wrong inside you and whether it's mild or severe—or, conversely, what's working well and shouldn't be messed with—how do you know what to change in order to boost your brain?

You don't need an onslaught of complicated, expensive, or invasive procedures just to get a snapshot of your overall health. If you came to see me as a client, whether you were an everyday person or a pro football player, the first thing I would recommend is basic blood work.

Common labs, like those ordered by a primary care doc during an annual physical, can indicate whether you have an underlying metabolic issue, a hormonal imbalance, or nutritional deficiency. Slight abnormalities in any of these areas are common—I've had low thyroid hormone and low vitamin D, for example (both of which I found out about by doing basic blood work). That's because metabolic, hormonal, and nutritional irregularities are often silent, rarely manifesting in acute symptoms, but causing vague side effects instead like fatigue, weight gain, and low mood.

Basic blood work can tell you and your doctor whether you have an underlying imbalance that may be secretly stealing your health. And while getting the labs is easy, there are a few steps you should know about.

Since there's no universal protocol for how often doctors should order labs for patients, you can't necessarily rely on your physician to take the lead. There's also no standard set of tests for every patient—it's up to your doctor to determine which ones he or she thinks you might need. But unless you walk into a medical clinic with a laundry list of symptoms, your doctor probably won't order you a thyroid panel, hormone panel, and C-reactive protein test. It's not that your doctor thinks these tests are unimportant, but many physicians don't order extra labs unless you ask.

Your first step, then, is to make an appointment with your doctor and ask him or her to order specific bloods tests for you. This is a normal and common procedure. While it can be intimidating at first to ask your doctor to do things rather than vice versa, remember, this is your brain, your body, and your health.

In my experience, physicians respond well when patients show they want to be proactive. I've found most doctors are more than happy to accommodate personal concerns, especially when these concerns are preventative—much different from a patient coming in and demanding a medication or quick fix to a condition better solved through lifestyle modifications.

One more note about basic blood work: Ask to receive a copy of the results so you can interpret, share with another practitioner if necessary, and save for your medical records. Many blood tests include a wide range of acceptable results considered "normal" yet not necessarily optimal. For example, the normal range for male testosterone is between 270 nanograms per deciliter (ng/dL) and 1,070 ng/dL. If you test at 275 ng/dL, you may be "normal" but not ideal. In this instance, you're functioning at suboptimal levels, and while your primary care doc might not detect it, a practitioner trained at maximizing personal health may be able to help.

Five Things to Know About Basic Blood Work

1. **YES, YOU CAN ASK YOUR PHYSICIAN TO ORDER BLOOD TESTS.** It's completely normal to ask your primary care doc to perform specific blood work to rule out a deficiency or an imbalance that can help optimize your health. Don't be intimidated—be confident.

2. **KNOW BEFORE YOU GO.** Don't get caught off guard by charges from your health insurance company after the fact. Check with your doctor's office or call your insurance company directly to find out which tests and panels are covered by your carrier.

3. **AVOID DIY.** In some states you can order your own blood tests online, but I wouldn't recommend it. Not only will you have to pay out of pocket for these tests, you'll also have to interpret your own results and/or rely on a computerized analysis that doesn't include a human perspective. If your primary care doc won't order a blood test, find another physician who will.

4. **SCHEDULE IN THE MORNING BEFORE YOU EAT OR DRINK.** Many labs require fasting, with no food or drink up to twelve hours before the test. Ask your doctor's office in advance

which labs require fasting, and be sure to adhere to their recommendations.

5. **LOOK FOR OPTIMAL, NOT NORMAL.** Explain to your doctor that you're trying to optimize your health and want to know any results that may be low, even if they're still considered within normal range. Ask for a copy of your blood work so you can assess the results yourself and get a second opinion if necessary.

Your Pregame Strategy: Eight Blood Tests to Help You Biohack Your Brain

1. **COMPREHENSIVE METABOLIC PANEL:** These are the most basic labs you can get, measuring blood sugar, electrolytes, and other compounds that can indicate proper fluid balance and blood filtration function.

   WHY YOU NEED IT: High blood sugar is toxic to the brain, interfering with cognitive function and significantly increasing the risk of Alzheimer's and other disorders. This test also shows whether you have enough electrolytes for proper fluid balance, optimal cerebral circulation, and other physical and cognitive functions.

   BE SURE TO ASK: Many doctors will order a basic metabolic panel unless you specify you want a comprehensive one, which also tests for certain blood proteins that can indicate kidney and liver function.

2. **FASTING BLOOD GLUCOSE:** Just as the name suggests, this lab specifically tests the amount of glucose, or blood sugar,

surging through your arteries and veins after a fast of eight hours.

WHY YOU NEED IT: A comprehensive metabolic panel will give you an indication of your blood sugar levels, and should be done following an eight- to twelve-hour fast. If this is not the case, be sure to do a fasting blood glucose test to determine whether you are diabetic or prediabetic. This test can be critical to preserving brain health: high blood sugar levels and insulin resistance increases the risk of Alzheimer's disease, which is sometimes referred to as type-3 diabetes.

3. **HEMOGLOBIN A1C:** Hemoglobin A1C measures how much sugar is attached to your red blood cells. This test is an assessment of your average blood glucose over the previous three months.

   WHY YOU NEED IT: This test will inform your doctor on whether you are diabetic or prediabetic and is helpful in the management of diabetes medications. I recommend multiple tests for these conditions because both are so prevalent and oftentimes go undiagnosed. Thirty million people in the United States are diabetic, and one in four don't know it, while 84 million Americans are prediabetic, with 90 percent of them in the dark about their condition.[1] For these reasons, the CDC recommends anyone over the age of forty-five get a hemoglobin A1C test, along with anyone under the age of forty-five who is overweight, doesn't exercise regularly, or has another risk factor for diabetes. Given that the majority of Americans are overweight and inactive, this test is a necessity for almost everyone.[2]

4. **LIPID PANEL:** Lipids are substances like fats or cholesterol that can't be dissolved in water. A lipid panel measures cholesterol specifically, including total cholesterol, triglycerides, high-density lipoprotein (HDL) cholesterol (the healthy kind),

and low-density lipoprotein (LDL) cholesterol (the unhealthy kind).

WHY YOU NEED IT: Elevated levels of triglycerides, LDL, and total cholesterol can clog arteries and block blood flow to the brain, cutting off the supply of oxygen and nutrients. High LDL levels may also fuel the development of Alzheimer's disease,[3] while having too many triglycerides in the blood can impair memory and executive function.[4] The good news is that having more HDL cholesterol may help prevent Alzheimer's and other neurodegenerative disorders, according to research.[5]

5. **C-REACTIVE PROTEIN TEST:** C-reactive protein (CRP) is a substance produced by your liver whenever your body is battling inflammation. This test measures CRP levels in your blood, telling doctors whether you're suffering from an unhealthy amount of internal inflammation.

   WHY YOU NEED IT: Inflammation minimizes cognitive function and increases the risk of nearly every illness and ailment. A CRP test may also indicate if you have a chronic inflammatory illness like rheumatoid arthritis.

   BE SURE TO ASK FOR: Opt for a high-sensitivity CRP test, which can indicate your risk of heart disease more effectively than the basic CRP test or a lipid panel.[6]

6. **VITAMIN D TEST:** Doctors have recently become aware of just how important this test is for the majority of patients. As the name implies, the lab measures whether you have enough vitamin D in your blood.

   WHY YOU NEED IT: Having low vitamin D can cause inflammation levels to skyrocket, leading to cognitive dysfunction and disorder, not to mention weight gain and an increased risk of diabetes and cancer. New research on vitamin D shows it may also stimulate the immune system to help clear amyloid plaque buildup that can cause dementia and Alzheimer's disease. Vitamin D has also been shown to regulate mood, as low levels are associated with depression. Despite its myriad benefits, 95 percent of all Americans don't meet the recommended daily intake.

7. **HORMONE PANEL:** With a different test for each gender, this panel checks whether you're producing enough sex hormones. The female hormone panel typically measures estrogen, progesterone, follicle-stimulating hormone (FSH), and testosterone/dehydroepiandrosterone (DHEA). A standard male hormone panel assesses testosterone, estradiol (a form of estrogen present in men), and DHEA, a steroid hormone that helps produce testosterone and estrogen.

   WHY YOU NEED IT: Many habits like diet, physical activity levels, sleep patterns, prescription drug use, and exposure to toxins in our food supply and the environment can interfere with our hormone levels. Having a hormonal imbalance, in turn, can wreak havoc on our brain, creating inflammation, stress, and cell damage while impairing memory, executive function, and mood. Hormonal imbalances are also a leading cause of fatigue, weight gain, sleep disturbances, libido dysfunction, and mood irregularities.

8. **THYROID PANEL:** This lab has become increasingly essential as doctors have learned more about just how much our thyroid hormones affect our overall health—and how many people, primarily women, have thyroid-hormone imbalances. This test measures how well your thyroid is functioning and whether you have the right amount of several different thyroid hormones, including triiodothyronine (T3), thyroxine (T4), and thyroid-stimulating hormone (TSH).

   WHY YOU NEED IT: If any of your thyroid-hormone levels are too low, it can thwart your memory, executive function, and ability to concentrate while increasing the risk of depression and other mood irregularities. Hypothyroidism, or having low thyroid-hormone levels, can also cause weight gain, fatigue, muscle and joint pain, and a host of other unpleasant symptoms. Conversely, hyperthyroidism, or having high thyroid-hormone levels can cause weight loss, rapid heartbeat, sweating, and irritability.

BE SURE TO ASK FOR: Tell your doctor you want the full thyroid panel, not just a TSH test, which is sometimes included in a standard hormone panel. The reason: your TSH levels can be normal even if your body doesn't produce enough T3 or T4.

Fire Chief Ken's Story

HOW BLOOD WORK HELPED HIM ADDRESS ADDICTION PROBLEMS WHILE OVERHAULING HIS BRAIN

Chief Ken, fifty-nine, had dedicated thirty-seven years of his life to the fire service before we met. Since my father had also been a firefighter, I already had a lot of compassion for the occupational exposures and hazards his brain and body had endured after decades on the job.

Given the prevalence of addiction among those in the fire service, I wasn't completely surprised when Chief Ken shared that he was attempting to manage a compulsion of his. He had reached out for my help to try to get a better understanding of the brain, which he hoped would help him to manage his addiction. At the time, he also had issues with dizziness, fatigue, poor coordination, and balance, and had gained weight since his early years in the service. He'd also recently been diagnosed with general anxiety disorder and was beginning to experience short-term memory loss. For someone who put his life on the line to save others, these symptoms were a wake-up call that he had to take action to save his own life.

My first recommendation was that the chief get a basic lab panel done. The results were illuminating. His BMI was 35, making him clinically obese, and he had high blood glucose, blood pressure, cholesterol, and triglycerides, along with low vitamin D. His lab results also revealed he had hypothyroidism, when the body doesn't produce enough thyroid hormones, which was contributing to his fatigue and weight gain.

Since he was overweight and always tired, the chief also went in for a sleep study, which revealed he had sleep apnea and needed a continuous positive airway pressure (CPAP) machine. When he tracked his current diet, we found out he was a stress eater, subsisting on fast food, sugar, and soda, often up until he went to bed.

At the same time, the chief entered an in-patient treatment center to address a gambling addiction. While he was receiving psychological help, we took steps to address his neurological symptoms. The chief's labs suggested his diet needed immediate overhaul, with his weight, cholesterol, triglycerides, and blood pressure all paying the price. The goal was to reduce his meat consumption, and I encouraged him to eat only organic or grass-fed cuts when he felt he had to, and to swap out processed junk for plant-based whole foods.

While we were making upgrades to his diet, the chief also started exercising regularly, eventually joining a nine-week running program and completing his first 5K race before his sixty-third birthday. He had also started tracking his steps, aiming to take at least six thousand per day and often topping out at ten thousand. Finally, with the help of his CPAP machine, he started averaging 7.5 hours of uninterrupted sleep per night, which he tracked using his Garmin watch.

Recently, the chief had another round of blood work, which showed his hard work and dedication are working. He is now 225 pounds, down from 395 pounds, without the help of beta-blockers, ACE inhibitors, statins, or other prescription drugs apart from thyroid medication. The chief has also knocked down his blood glucose, blood pressure, cholesterol, and triglycerides to normal range.

Now that his weight has stabilized, the chief also no longer needs the CPAP machine. While he's still working to get his BMI into a healthy range, he's better able to manage any problems that arise and has an arsenal of strategies to stay on track. He lives a more empowered life and feels stronger, happier, and healthier.

KRISTEN'S TIP: Basic blood work can be the motivation that inspires you to change your health and life. For the chief and many others, it provides a series of benchmarks to aspire to. Just like seeing your weight drop on a scale, getting your lab numbers within a healthy range is not only rewarding but also transformational.

Getting Started: Four Ways to Turn Goals into Action

Many people start out with amazing goals and intentions. You want to change your brain and you want to do it now, so you

The Single Test That Can Save Your Brain

Getting a hearing test may be the single most important assessment for your brain. According to the *New York Times,* "hearing loss in the largest modifiable risk factor for developing dementia, exceeding that of smoking, high blood pressure, lack of exercise, and social isolation."[7] Even a slight hearing loss still considered "normal" can minimize brain performance, limiting your ability to think clearly, be rational, and remember details.[8]

When we can't hear as well as we should, it forces the brain to work harder, preventing it from completing more essential tasks. Hearing problems also socially isolate people, increasing the risk of dementia. Untreated hearing loss can actually drive up dementia risk by 50 percent in five years, also boosting the likelihood of depression by 40 percent in the same time.[9]

The good news: getting a hearing test is easy. Talk with your primary care doctor, who may recommend an audiologist. If you already have hearing loss, consider using a hearing aid. While some don't like how these devices look, feel, or sound, wearing one can make the difference between a healthy, functioning brain and cognitive decline and dementia. In the meantime, reduce your risk of hearing loss by wearing ear plugs or noise-cancelling headphones during loud events like concerts and construction work, and keeping the volume low while listening to music or podcasts through headphones.

commit to overhauling your diet, exercising daily, sleeping eight hours a night, meditating in the morning, going to yoga after work, and losing that extra weight you've been carrying for years.

This is exactly the ambitious, optimistic outlook you need to start making changes that can transform your cognitive power and performance, but to sustain these changes over time, you'll likely need a little help. Otherwise, the odds aren't in your favor that you'll be able to maintain them long enough to impact your physical, mental, or cognitive health.

Conquering new lifestyle habits has nothing to do with your degree of determination or self-discipline. Ninety-five percent of diets don't work, according to research, with most people regaining any weight they lose in a matter of months or even weeks.[10] Similarly, only 8 percent of people who make New Year's resolutions actually accomplish them.[11] Eight percent!

If it's not a matter of dedication and self-discipline, what do we need in order to sustain new habits? Here are four ways to turn your goals into reality and make new habits an easy, enjoyable part of your lifestyle rather than a continual effort.

1. Start Small

Studies show people who set realistic, incremental goals are more likely to accomplish them than those who attempt to overhaul their diet, exercise, sleep, and other habits from the beginning.[12]

For example, if you eat sugary cereal or bagels for breakfast, like sandwiches or burritos for lunch, and order pizza or takeout for dinner, with some coffee, sports drinks, and/or alcohol in between, it's going to be difficult to swap out all processed foods, caffeine, booze, and sugary drinks for seafood, legumes, and fruits and vegetables. You'll likely feel deprived and suffer cravings, which are easy enough to overcome, but you won't have had the time or experience to develop satisfying alternatives to the unhealthy foods you're used to consuming.

In this instance, instead of giving up everything all at once, start by removing most, then all processed foods from your diet. Next, you can work on reducing daily caffeine consumption by weaning yourself down to one cup, then eventually a half cup of coffee. When you feel you have your consumption of processed foods and excess caffeine under control, strive to reduce how much alcohol you drink and cut other forms of added sugar from your diet.

Something else to consider: you may not want to change your diet, fitness routine, hydration schedule, sleep regimen, stress-management protocol, and supplement program all at once. I've learned by coaching clients that some like to implement multiple modifications right away while others feel overwhelmed by the idea of changing two things at the same time.

No matter which camp describes you, I'd encourage you to commit to changing just one habit per week. Then, every subsequent week, strive to maintain the lifestyle modifications you've already changed as you continue to adopt a new habit. At the end of ten weeks, you'll likely find you've been able to make ten small, meaningful changes that, collectively, can have a significant impact on your brain and overall health.

2. Track It

Monitoring the number of calories you eat, steps you take, ounces you drink, and/or hours you sleep makes it more likely you'll strive to see healthy numbers, which will help you meet your goals. Tracking through apps or wearable devices also provides instant feedback, allowing you to make micro-adjustments in real time. In some instances, tracking also lets you share your data with a group of friends or a like-minded online community, adding support, motivation, and inspiration.

For all these reasons, people who track their habits are more likely to hit their health goals than those who don't self-monitor, according to research.[13] For example, people who log their daily calories lose more weight than those who don't,[14] while those who monitor their physical activity levels get to the gym more often—and actually enjoy the process—than those who don't track.[15]

I recommend all my clients track any new health habits for a minimum of twelve weeks. Three months will expose any patterns that may develop and help you make changes if something isn't working. For example, if you start tracking your calories, you may discover you're eating more than you thought because you're mindlessly snacking while preparing dinner. One solution might be to enact a "no-nibble rule" while cooking or ask your spouse to prepare dinner for a week to see what effect it has on your weekly caloric intake.

Personally, I track nearly every aspect of my health, including the six metrics I've outlined below. This allows me to make adjustments in real time, preventing or curtailing derailments to my health. For example, I recently discovered by tracking calories that I was consuming too much sugar. I was buying dried mangoes in bulk from Whole Foods, and while the item is relatively healthy, dried fruit, as we know, is loaded with sugar and calories. A little here and there isn't a problem, but I realized I was mindlessly snacking on the dried mangos on my way home from work and consuming hundreds of calories and dozens of sugar grams as a result. When I stopped, my blood sugar stabilized, and I had more energy and less of an appetite on the days I went food shopping.

Which habits should you monitor? That answer depends partially on your goals, but I would start by tracking the following six metrics here. That said, if you drink water like a fish, you may not need to track your hydration. Or if you do yoga daily to keep your stress in check, you won't need to monitor meditation. It's up to you, but the more you track, the more success you'll have, I guarantee.

1. **WEIGHT**. One of the best ways to optimize your brain power and performance is to get to and maintain a healthy weight. But what weight is considered healthy? One of the best ways to determine healthy weight is to assess your Body Mass Index (BMI). While some trainers may argue fat percentage is a more accurate tool, the CDC and other national experts say

BMI provides a better snapshot of your overall health risks as it relates to body weight.[16] However, if you have access to a professional who can interpret your fat percentage, you can use that as a basis for your weight goals.

Otherwise, you can easily learn your BMI by using a BMI calculator online and imputing your height and weight—I recommend accessing the National Institutes of Health's calculator at https://www.nhlbi.nih.gov/health/educational/lose_wt/BMI/bmicalc.htm.

A BMI of greater than 30 indicates you're likely obese and should see a physician or another health care practitioner who can create a targeted weight-loss program for you. A BMI of 25 to 29.9 means you're overweight and need to shed pounds. Following my Better Brain Diet can help accomplish that goal if you combine the program with a low-calorie diet (see the next metric on page 212).

If your BMI is between 18.5 and 24.9, then you're at a healthy weight—congratulations! A BMI of less than 18.5 indicates you're likely underweight and may need to see a doctor to make sure you're getting enough nutrients to support your brain and body.

Once you know your BMI and whether you need to lose or maintain weight, I recommend that you start hopping on a scale daily to monitor whether you're losing to reach your goal or maintaining. While your weight will fluctuate a few pounds day to day, weighing yourself daily makes it a habit and also allows you to make micro-adjustments to your calorie consumption and activity levels accordingly. This, in turn, helps prevent those three extra pounds from turning into thirty—a more common occurrence as we get older and our metabolism begins to slow down. Weighing yourself regularly can also show you whether you're gaining weight for reasons other than what you eat or how you exercise, like if you're taking certain medications, sleeping too little, or not managing your stress, all of which can cause the body to hang on to fat.

For these reasons, studies show that people who weigh themselves daily are more likely to lose weight than those who don't, regardless of which diet they follow or how much they exercise.[17] For consistent results, weigh yourself the same time every day, preferably first thing in the morning.

2. **CALORIES**. This is one of the most potentially eye-opening and life-changing metrics you can track. Most people have no idea how many calories they consume on a daily basis. But when they start tracking, they're shocked to discover they're actually eating and drinking way too many calories for their weight and height. It doesn't matter if you eat mostly protein and fat with zero carbs—if you consume too many calories, you will gain weight.

   Before you begin tracking, you'll want to figure out how many calories you actually need on a daily basis to maintain your weight and be healthy. Most calorie-counting apps include a calorie calculator that can compute your daily energy requirements based on your age, gender, height, weight, and activity level. If the calorie-tracking app you like doesn't include a calorie calculator, you can easily find one online. One qualifier: Unless you're extremely active, I recommend you enter "sedentary" or "light" in the physical activity prompt, as most people overestimate the effect exercise has on calorie expenditure.

   The next step, if you haven't already, is to find a calorie-counting app that will let you log everything that you eat and drink and figure out your daily calorie consumption for you—FitBit, Lose It!, and MyFitnessPal are all popular options. Be sure to enter every nibble, sip, and bite, even if it's just a small sample cup of a smoothie from your favorite juice bar or the peanut butter off the knife from the sandwich you make—these calories, especially from energy-dense foods, can add up quickly and tremendously. Use the counter to figure out how many calories you're eating and drinking daily and whether you need to consume less to maintain or lose weight, depending on your BMI and daily caloric goal.

3. **EXERCISE.** You're more likely to do anything when you log it—and that's especially true of exercise. Tracking your workouts helps you stay accountable, also providing a huge sense of satisfaction when you can visually see your progress, day after day, week after week. For this reason, people who track their workouts increase their activity levels far more than those who don't, according to research.[18]

   You can track your workouts by writing them down on a wall calendar or in an old-fashioned journal, which may be more gratifying if you find it rewarding to write down your achievements rather than type them into an app and want to be able to easily measure your progress from a visual standpoint.

   On the other hand, hundreds of people have had tremendous success using fitness apps and wearable devices like FitBits. The benefit of apps is that many, like the step counters found standard in many smartphones, track your overall activity, giving you extra incentive to move more, whether it's through a traditional workout or simply by walking more. Some apps like Strava and FitBit (when paired with a device) also allow you to share your workouts with friends, your personal trainer, and/or an online fitness community, which can provide extra motivation, support, and accountability. Finally, some apps like FitBit let you track your weight, calories, hydration, sleep, and physical activity in one place, allowing you to consolidate your self-monitoring.

   Personally, I'm obsessed with the iPhone app Stepz, which tracks daily steps, total mileage, and calories burned. The app challenges users to meet a daily goal of ten thousand steps—widely cited as necessary for good health—and congratulates you when you reach that goal or encourages you to walk more by flashing different colors (orange if you're close, red if you really need to pick it up).

Before I discovered Stepz, I was clueless about how much activity I got outside of my daily dedicated workouts. Why are steps important if you still work out every day? Research shows that contained exercise can't undo all the deleterious effects of sitting all day and that the healthiest people are also active outside the gym.[19] When I started getting more steps, I noticed I had more energy and mental clarity, and felt more limber throughout the day.

4. **HYDRATION.** There are countless low- or no-cost apps that will calculate your fluid requirements based on your height, weight, and gender, allowing you then to input your daily intake to help you meet your hydration goals. Some of these apps send alerts signaling you when to drink, while others like Waterlogged let you customize the size of your glass or bottle to better tally how much you're drinking. One clever app called Plant Nanny even includes a cute animated flower that flourishes the more you drink. If you prefer, you can also write down how many ounces you drink in a hydration journal, using the Institute of Medicine's standard that men consume 3.7 liters and women 2.7 liters of water as your daily goal.

Before I started tracking my hydration years ago, I had no idea how little water I was actually drinking. The process of logging my daily fluid intake was both illuminating and horrifying, as I realized I wasn't consuming nearly enough to keep my brain hydrated and healthy. That was when I started carrying thirty-two-ounce stainless steel bottles filled with purified water everywhere I went, making sure I finished three full bottles before I went to bed. Today, I would still miss my hydration goal unless I track my fluid intake or ensure I maintain my three-bottle routine.

5. **SLEEP.** Most people overestimate the number of hours they sleep. If you track your sleep though, you'll get hard data of how much you're actually sleeping and whether you need to make changes to your routine in order to improve your brain health and function.

   Tracking your ZZZs can also help you identify whether any symptoms you may be experiencing are due to poor sleep. While it can be easy to blame symptoms like daytime fatigue, brain fog, poor memory, increased appetite, weight gain, lack of motivation, anxiety, and depression on a host of other problems, sleeping a full eight hours may be all your brain and body really need.

   Sleep trackers, including smartphone apps and wearable devices, can estimate your sleep quantity, quality, and time spent in certain phases like deep sleep and the dream state known as rapid eye movement (REM). Some can even time an alarm to rouse you when you're not in deep sleep, which will limit grogginess and make it easier for you to wake up. Popular options include wearable devices like FitBit Versa and low-cost apps like SleepScore—my personal favorite.

   Keep in mind that sleep trackers can't take the place of a sleep apnea test. If you suspect you might have sleep apnea (see page 151), make an appointment immediately to see your physician.

6. **MEDITATION.** If you choose to use meditation to lower your stress (see page 153 for why you should consider adopting the practice), you can track your progress, get guidance for your practice, and even learn what's happening inside your

brain with a number of downloadable apps and wearable devices.

Apps like the Mindfulness App and Sattva offer guided sessions while allowing you to log how often you practice. This helps increase accountability and also lets you identify trends that might develop from your practice, like whether you feel less stressed and more focused during the weeks when you meditate more often.

Wearable meditation sensors like Muse take the concept of a downloadable app and amplify its data-tracking abilities tenfold. I love Muse because it provides real-time feedback on what's happening inside your brain as you meditate. Muse monitors your brain waves, signaling you with the sound of an approaching storm when you need to calm your thoughts while using peaceful weather and the sound of chirping birds when your mind is perfectly relaxed. The device also pairs with a smartphone so you can track your progress and cognitive statistics, helping you stay focused, motivated, and meditating regularly.

Muse isn't the only company to offer a meditation sensor, although I recommend it because it relies on fundamental principles in neuroscience. Be warned, however, that meditation sensors can be expensive, costing as much as $200 for a basic model.

3. Find a Coach or an Accountability Partner

Having someone in your life who can hold you accountable for reaching your goals will increase your chances of meeting them by as much as 65 percent, according to research. And if you can meet regularly with that person, your chances of success go up to as much as 95 percent.[20] An accountability partner can be a spouse, close friend, colleague, or trained professional like a personal trainer, nutritionist, therapist, or cognitive coach. Checking in with your accountability partner doesn't have to take time or effort, either. It can be as easy as a five-minute phone call during which you share your tracking data for a specific target or set of targets. One reason my clients succeed in accomplishing their health goals is because they tell me they feel like they have a constant coach and cheerleader in me who not only helps them find the right path but also supports and encourages them along their way to a better brain.

Why You Should Track Your Blood Pressure

Approximately 11 million Americans have high blood pressure without knowing it, according to the CDC.[21] High blood pressure, or hypertension, usually doesn't carry acute symptoms—that's why it's often called the "silent killer." When your blood pressure is too high, the resulting force on your vessels can damage arteries and prevent blood from getting to your brain, not to mention the rest of your body, while putting a massive strain on your heart.

Keep abreast of your blood pressure by downloading a free app like SmartBP or Cardio Journal that can measure this metric and let you know when it might be time to see a physician. You can also use a wearable device like an armband or a wristwatch, which costs a little money but is far more accurate. I use an arm cuff from Omron to track my blood pressure and heart rate daily—the device also stores my previous data so I can compare results. If you already have or suspect you have unhealthy blood pressure, don't wait—see your physician. Similarly, if you already have high blood pressure, follow your doctor's advice on how to best treat and monitor it.

4. Enjoy It

Consider creating a healthy incentive for yourself, like a month's delivery of organic fruits and veggies or a vacation somewhere relaxing, if you lose ten pounds, meditate daily for two months, or successfully give up alcohol and coffee. Find and embrace the workouts that you enjoy most, make a pact with a friend to hit a health goal or go for long walks together—whatever brings you joy. Through it all, just remember: you're on an incredible journey to be the smartest, healthiest, and happiest person you can be, and that's worth celebrating at every step.

The NFL Story

HOW THIS FOOTBALL GREAT CHANGED THE PLAYBOOK FROM X'S AND O'S TO STARS OF HOPE

I first told you about Ed White in chapter six and how giving up coffee at age sixty-two helped this former Minnesota Vikings offensive guard change his brain (see page 142). But there's another reason I love Ed's story. He recently created a successful way to track habits that has overhauled his health and turned an unfortunate situation into something positive.

Two years ago, Ed was diagnosed with Alzheimer's, a fate that changed his life but not his outlook. Instead of succumbing to the disease, he's decided to tackle it like any other threat on the football field, going back to the high-performance health habits he learned during our sessions together. He may not have optimal cognitive health now, but he still wants to make his brain the best it can be.

A year ago, Ed began tracking five metrics—his weight, calorie intake, sleep, steps, and number of hours he intermittently fasts—and writing down his results in a journal. At the end of each day, he scores himself on each metric, giving himself a star when he feels he's achieved his goal. If he misses the mark, there's no punishment that might demoralize him. At the end of the month, he tallies up how many five-star days he has. The following month, he tries to maintain or exceed his five-star days.

Ed is compassionate with himself when he awards stars. He'll give himself a star if he loses weight, but also if he maintains it. He uses MyFitnessPal to track his calories and gives himself a star if he doesn't exceed his daily maximum. He fasts between dinner and his first meal the next day (learn more about intermittent fasting on page 71), awarding himself a star if he fasts sixteen hours total. He also tracks his sleep with Fitbit, giving himself a star when the app ranks his sleep as "good." Ed also uses FitBit to log his steps: His daily goal is ten thousand, but he'll give himself a star if he exceeds five thousand.

Since Ed started tracking his habits again, he's lost seventy-seven pounds, improved his sleep, overcome joint pain, and increased his mental alertness and acuity. When he skips a few days of monitoring, he tells me he dips back into overeating, poor sleep, and little exercise. But he doesn't beat himself up about it. Instead, he acknowledges the miss and gets back on track, literally. More important, he likes monitoring his metrics: it gives him a sense of control and has turned his goals into a game.

Ed enjoys monitoring to such an extent that he recently decided to track five more metrics: hydration, supplements, how long he plays brain games every day, his green-juice consumption, and his blood pressure. He now gives himself stars whenever he drinks enough water, takes all his supplements, plays brain games for twenty minutes, drinks a green juice, and keeps his blood pressure in the normal range, according to his Fitbit. This means Ed can now log a ten-star day.

One year later, he says he feels stronger and sharper, despite his diagnosis.

KRISTEN'S TIP: Self-monitoring can be motivating and empowering. It can also be a form of self-care and self-compassion. For Ed, tracking allows him to try to be better every day than he was the day before.

EPILOGUE

Biohacking Your Brain in the Twenty-First Century

Congratulations—you now have everything you need to biohack your brain! If you leverage every tool outlined in this book—diet, exercise, hydration, supplements, stress regulation, optimism, and cognitive training—you can create the healthiest mind within your immediate power.

For those of you who would like to take your biohacking to the next level, there are a few different options I would encourage you to explore. After all, technology has accomplished amazing feats in the field of cognitive health and performance, and there are certainly new and emerging tools being developed to evaluate and treat your brain.

I will say that the options I list are not necessarily cheap or easily accessible, but they are helpful if you feel you need an additional boost beyond all the tools we've already discussed.

Here are four things to pursue if you want the healthiest, most high-powered brain possible.

NEUROFEEDBACK: This is one of the most promising interventions for the brain, given that it improves the stability and efficiency of the network connections, helping to strengthen your cognitive proficiency and power. If you're familiar with biofeedback—a common therapy used to help control the body's physical responses, like heart rate, blood pressure, and muscle tension—then you have a good understanding of how neurofeedback works. Neurofeedback is simply biofeedback for the brain, using electroencephalography (EEG) to measure the electrical activity inside your brain. With EEG, electrode sensors are placed on the scalp that communicate brain wave activity in real time to a computer. Those results are then interpreted by a clinician who helps guide you to moderate that activity in myriad ways.

Neurofeedback has a profound effect, helping rewire neural pathways to increase communication between different areas of the brain that can help make the mind more efficient, improving cognitive ability, creativity, and sustained attention. The therapy has been used to target dysfunctional areas of the brain and alleviate symptoms associated with chronic pain, depression, anxiety, trauma, insomnia, headaches, and other cognitive ailments. While most neurofeedback protocols require multiple sessions, even one hour of neurofeedback is all it takes to begin to see improvements in cognitive communication and strengthen neural pathways, according to research.[1]

Personally, I've had great success with neurofeedback, having used it myself and applying it in our clinical research. With our NFL players, we used neurofeedback to help strengthen connections in the brain that had been damaged by head trauma. We also used the therapy to address individual issues like ADHD, anxiety, depression, and insomnia. The treatment is drug free, with no side effects, yet what I love most about it is that you're not just putting a Band-Aid over the problem but actually retraining your brain to function more effectively for life.

Neurofeedback clinics are easy to find across the country. Speak with your primary care physician for a recommendation.

Should You Get a Brain Scan?

You've heard a lot about brain imaging throughout this book and how it's changed the lives and minds of many with certain neurological issues. But does that mean you should rush out to get a brain scan? Not necessarily. The first step is to speak with your primary care doctor or a neurologist, as only he or she can prescribe brain imaging. There are a variety of imaging options, and each serves a unique purpose in telling clinicians something different about your brain. Some, including single-photon emission computed tomography (SPECT), expose you to a small amount of radiation, which clears the body in a day or two, but it still pays to be careful what you expose your most precious organ to.

If you're interested in brain imaging and don't have a neurological condition, I recommend you ask your doctor about qEEG, short for quantitative electroencephalography, which can measure the electrical activity inside your brain. A qEEG test is noninvasive and doesn't use radiation, but can show a clinician just how efficiently your brain works, revealing which areas may be overactive or where neural connectivity may be weak. This, in turn, can help doctors recommend a protocol to optimize your cognitive function and help you better address mental or mood problems.

Ask your primary care physician about qEEG testing and the best places to have it done. A qEEG test usually costs a few hundred dollars—less expensive than other forms of brain imaging—although is not usually covered by health insurance unless it's necessary for a specific medical condition.

HYPERBARIC OXYGEN THERAPY: In chapter four, I told you that exercise was the most effective way to boost cerebral circulation—unless, that is, you have access to a hyperbaric oxygen therapy chamber. Hyperbaric oxygen therapy (HBOT) allows you to breathe pure oxygen in a small room or chamber that's been pressurized to be three times greater than normal air pressure. Under these conditions, your lungs can take in more oxygen, increasing blood delivery—and the oxygen and nutrients it carries—to your precious brain.

HBOT is primarily prescribed as an off-label use to repair cognitive damage in those who have suffered concussions, hard hits, or other traumatic brain injuries. For example, we used HBOT during our research with NFL players, who were able to make incredible improvements in their cerebral circulation and restore some of the blood-flow deficits visible on their SPECT scans. Research has also shown that those with Alzheimer's and other forms of dementia may be able to increase cognition and mitigate symptoms.[2] More studies are needed, however, before the therapy is widely prescribed for those without cognitive damage.

If you're interested in HBOT, speak with your doctor about writing you a prescription and referring you to an appropriate clinic. The therapy isn't appropriate for everyone, and there are some mild risks that should be discussed with a physician before use. Note too that the therapy often takes multiple sessions to impart lasting effects and can be expensive. Some health-insurance providers may cover the treatment if it's for a medically indicated use.

FLOTATION TANK: I absolutely love flotation tanks. I wish I owned one, because I'd use it every day to help restore my mind. Flotation tanks provide total sensory deprivation, as you float in salt water the same temperature as your skin, with no light or sound to distract you—essentially, it's a way to experience complete relaxation. The effect is brain-changing, instantly reducing stress, anxiety, depression, and even physical pain, according to research.[3] The therapy has also been shown to lower blood pressure and the stress hormone cortisol,[4] producing a euphoric state in many that helps counter the detriments of stress exposure.[5] Over time, float therapy has also been shown to treat anxiety,[6] addiction,[7] fibromyalgia,[8] and other neurocognitive or physical disorders.

I consider float therapy to be the ultimate in self-care and one of the best ways to mitigate individual stress. Look online to find a float-therapy center near you—many spas also offer the treatment. Individual sessions range in price, but some clinics offer monthly memberships for a reduced price.

HYPNOTHERAPY: Hypnosis isn't high-tech—it's been practiced for centuries—but the therapy now has a growing body of modern-day research to show it's an amazingly effective way to lower stress, limit negativity, and help treat trauma that can interfere with cognitive function and performance. Studies also show hypnosis can improve focus[9] and treat numerous conditions, including insomnia, chronic pain, tension headaches and migraines, irritable-bowel syndrome, addiction, and phobias. I've seen hypnosis help patients address a variety of issues from nicotine addiction to food cravings while retraining people's brains to be more optimistic and open to better health and healing.

Look for a licensed psychologist, doctor, or mental-health counselor who is also certified in the field of hypnotherapy. If you're treating a certain condition like anxiety, stress, addiction, or cravings, it may take several sessions to be effective. Call your health-insurance carrier before scheduling an appointment, as some will cover a percentage of the cost if you see an in-network provider.

Ultimately, all of these are just bonus suggestions in your quest to biohack your brain. The most important tools are the ones you already have—the motivation to take control of your health, and the knowledge to be able to do so. You don't have to apply everything you've learned at once. The beauty of biohacking your brain is that this is *your* journey—you can adopt different practices over time as you discover what works best or the ways to improve your own cognitive power and performance that you enjoy the most.

As you do so, remember there are millions on a similar quest to boost brain health—you are not alone. All over the world, many have chosen to empower themselves with the know-how to protect, preserve, and sharpen everything our amazing brain can do for us. After all, it's a gift to be able to think, act, and love. Cherish the gift for yourself, and also share it with others. Perhaps one of the best ways to biohack the brain is to open your heart, increase the love, and help others around you create a smarter, happier, and healthier life, too.

AFTERWORD

Finding Love and Happiness in a Post-Coronavirus World

Brain health is essential to overall well-being now more than ever. The coronavirus outbreak heightened the need we all have to prioritize every aspect of our health—not just our physical welfare but also our psychological well-being. The COVID-19 pandemic brought unprecedented levels of fear, anxiety, and stress to millions worldwide and inflicted unbelievable mental and emotional trauma, as thousands grappled with the loss of loved ones. At this time, we still don't know the psychological toll the outbreak will have on our collective well-being and, more profoundly, our individual psyches.

But with the help of *Biohack Your Brain,* you can make a conscious choice to fortify and heal yourself—body, mind, and spirit. Everything you've learned throughout this book will help you grow stronger, fitter, and more emotionally resilient to the kind of threat the pandemic presented. More importantly, this book includes the tools we all need right now to begin to mend the psychological trauma every single one of us faced during the coronavirus outbreak—and that many of us continue to endure on a regular basis. Focusing on your cognitive health now will also increase your psychological resilience to withstand future trauma, God forbid another outbreak threatens our world again.

For many, the disruptions to daily life presented by the coronavirus crisis have made it difficult to envision a bright future, but I want to tell you that you *can* be healthy and happy. It's a conscious choice you make by focusing on how you eat, move your body, train your brain, fortify your mind, engage in your relationships, communicate with those around you, consume information, and nurture yourself.

But it's not just about choosing nutrient dense foods and getting more exercise. You also need to decide how you focus your mind, choosing positive thoughts and love over negative thoughts and fear. Brain imaging shows that people who focus on optimism and love rather than pessimism and hate are better equipped to withstand the anxiety and fear we all face—two overriding emotions for many during the viral outbreak.

How do you choose optimism and love? Chapters seven and eight outline many strategies, but the best way, in short, is to take time to self-reflect, show gratitude, and practice acts of kindness. These three things can shift your mind from fear and anxiety to one of joy and calm.

Whether you choose to practice meditation, yoga, or breath work, taking time to self-reflect will calm your mind, reduce fear, and make you feel more hopeful about your life and the world around you. Expressing gratitude and reminding yourself of all the reasons you have to feel fortunate—starting with the fact that you're alive on this beautiful Earth—will infuse your mind with positivity, increase your emotional resilience, and enhance your psychological well-being. Finally, helping others in any little way you can, whether it's saying a kind word to a neighbor or listening with compassion to other people's problems, activates the brain's reward centers to stimulate greater feelings of happiness and joy.

Finding the coping strategies that help you manage your fear and anxiety can change your brain's activity in positive ways. It can also give you a deeper, more meaningful connection with yourself, your community, and even humanity as a whole. If there's one blessing from the coronavirus crisis, it's cultivating this gift of love.

ACKNOWLEDGMENTS

With gratitude.

I became fascinated with neuroscience and studying the brain in 1998 when I started my graduate education at UCLA. This passion grew exponentially after spending time with a very special group of people in academia who have influenced my scientific career in meaningful ways. I am eternally grateful for their unwavering support and for the many thoughtful and reflective conversations over the years. This includes Dr. Barney Schlinger, my first academic mentor, who extended an opportunity to work in his neuroendocrinology lab. Thank you for believing in me as a young scientist and for training me in the art of scientific writing for scholarly journals. I am so fortunate to have been trained by two exceptional mentors during graduate school: Dr. Felix Schweizer and Dr. Stefan Pulst, whose expertise in neurophysiology and genetics, respectively, helped to broaden my training and laboratory skills across multiple disciplines in neuroscience. I would also love to acknowledge all of the bright fellows, post docs, graduate students, and laboratory assistants for all of the thoughtful discussions: You made my graduate experience at UCLA and Cedars-Sinai Medical Center so rewarding. I honor and thank my dear friend and mentor Dr. Daniel Amen, who provided the opportunity for me to translate basic neuroscience into the clinical setting, using functional neuroimaging to study the efficacy of therapeutic approaches in psychiatry. Thank you for teaching me how to communicate neuroscience information in a way that is connected, compassionate, and relatable.

I would also like to acknowledge a group of colleagues whom I hold in high esteem for their wisdom, knowledge, and incredible support along the way, including Dr. Jack Feldman, Dr. Keith Black, Dr. Robert Thatcher, Dr. William Mobley, Dr. Mark Gordon, and Dr. Valentin Rushty.

I also want to extend gratitude to those in my inner circle of support, for collectively, each of you contributed to my life in a way that made the publication of this book possible. This starts with my incredible partners in publishing at HarperCollins, SVP and Director of Creative Development Lisa Sharkey, whom I knew on meeting truly understood why I am so passionate about spreading the message of brain health to the world, and my editor, Anna Montague, who has been so unbelievably helpful, patient, and kind throughout the entire book publishing process. I also want to thank the rest of the talented team assisting me, including Maddie Pillari, Maureen Cole, Kaitlin Harri, and Christina Joell, and a very special thank-you to my friend and cowriter Sarah Toland, for your sharp intellect and wit. Thank you to my wonderful agents, Babette Perry of Innovative Artists and Ian Kleinert of Projector Media, who championed this work. I am profoundly grateful for your belief in me. And most important, thank you to my family, Barry Isaacson, Bill and Patricia Cegles, Paul and Rose Cegles, Judith Pearson, Samantha and Tony Solimine, and Bob and Suzanne Pearson, for your unconditional love throughout this process.

I also wish to acknowledge the love of my life, Mark, whose commitment and encouragement uplifts my spirit and makes every day a blessing. Thank you for your belief in me and for being an endless source of strength and support. And thank you to our ever-present, faithful Teddy Roosevelt terrier, Oscar, who brings comfort, love, and joy to our life and provides a type of bonding that feels unique to animals who are rescued.

GLOSSARY OF BRAIN-RELATED ACRONYMS

ALA: alpha-linolenic acid—a type of essential fatty acid and omega-3 fatty acid that is found in nuts, canola oil, flaxseed, and other plant foods

ALC: Acetyl-l-carnitine—the supplemental form of the amino acid carnitine, which helps brain cells produce energy

BDNF: brain derived neurotrophic factor—a protein that helps stimulate neurogenesis and increase feelings of optimism and good mood

BMI: body mass index—a value derived from a person's height and weight that can indicate whether someone needs to lose, gain, or maintain weight for optimal health

BPA: bisphenol-A—an industrial chemical found in some plastics and other products that can be harmful to physical and cognitive health

CBT: cognitive behavioral therapy—a type of psychotherapy that helps address underlying thought and behavior patterns that can be harmful to cognitive, mental, and emotional health

COQ10: coenzyme Q10—an antioxidant often taken as a supplement that helps protect cells from damage and aids in metabolism

CPAP: continuous positive airway pressure—a form of therapy that helps treat patients suffering from obstructive sleep apnea

CRP: C-reactive protein—a substance produced by the liver in response to inflammation

CTE: chronic traumatic encephalopathy—a progressive degenerative brain disease found in people with a history of repetitive brain trauma, most notably in football players and military vets

DHA: docosahexaenoic acid—a type of essential fatty acid and one of two marine omega-3 fatty acids that is found primarily in seafood, meat, and some plant-based sources like seaweed and algae

DHEA: dehydroepiandrosterone—a steroid hormone that helps produce testosterone and estrogen

EEG: electroencephalography—a noninvasive procedure that measures the electrical activity inside the brain without the use of radiation

EFAs: essential fatty acids—a type of fat critical to physical and cognitive function that the body does not manufacture, necessitating consumption through diet and/or supplements

EGCG: epigallocatechin gallate—an antioxidant found primarily in green tea that helps to protect cells from oxidative stress and lowers inflammation

EMFs: electromagnetic fields—invisible areas of energy that contain low-level radiation, emitted by electrical equipment and wireless transmitting devices like cell phones, computers, WiFi networks, and microwave ovens

EPA: eicosapentaenoic acid—a type of essential fatty acid and one of the two marine omega-3 fatty acids that is found primarily in seafood

GABA: gamma aminobutyric acid—a naturally occurring amino acid and neurotransmitter often taken as a supplement to help reduce anxiety and promote sleepiness

HBOT: hyperbaric oxygen therapy—a type of therapy in which a person breathes oxygen-rich air in a highly pressurized chamber, making it possible for the lungs to absorb more oxygen

HDL: high-density lipoproteins—frequently called the "good" cholesterol, HDL helps carry cholesterol to the liver, where it can be removed from the body

HIIT: high-intensity interval training—a type of anerobic training that includes fast-paced intervals to help increase fat metabolism and improve pulmonary and cardiovascular function

IQ: intelligence quotient—a test created by psychologists that assesses academic progress based on a series of cognitive tests

LCTs: long-chain triglycerides—a type of fat found in most fatty foods like butter, vegetable oil, meat, and dairy

LDL: low-density lipoproteins—frequently called the "bad" cholesterol, LDL in high levels can lead to a buildup of cholesterol in the bloodstream

MCTs: medium-chain triglycerides—a type of fat found primarily in coconut and palm kernel oils that is shorter in structure and more easily metabolized by the body than long-chain triglycerides

MIND: Mediterranean-DASH Intervention for Neurodegenerative Delay—a type of diet developed by researchers at Rush University Medical Center to help reduce the risk of neurodegenerative disease and cognitive decline

NAC: N-acetylcysteine—the supplemental form of the amino acid cysteine and a powerful antioxidant that can help in supporting a balanced mood

NSAID: non-steroidal anti-inflammatory drugs—a class of medicine that includes aspirin and ibuprofen often taken to help reduce pain

NVT: neurovisual training—a type of cognitive training that uses simulators, computer screens, and virtual reality headsets to challenge eye movements and increase overall optical skills

PCBs: polychlorinated biphenyls—a group of industrial chemicals harmful to physical and cognitive health, often found in seafood products

PS: phosphatidylserine—a fatty substance often taken as a supplement that is responsible for healthy nerve function

QEEG: quantitative electroencephalography—an analysis of the brain, also known as "brain mapping," that uses electroencephalography to determine electrical activity inside the brain

SPECT: single-photon emission computed tomography—a functional nuclear imaging technique that allows doctors to analyze blood flow to the brain

TSH: thyroid-stimulating hormone—a hormone measured by a TSH test that can help indicate how well the body's thyroid is functioning

NOTES

Publication Abbreviations

AAPS J—The AAPS Journal

ACSMs Health Fit J—American College of Sports Medicine's Health & Fitness Journal

Acta Neurol Taiwan—Acta Neurologica Taiwanica

Adv Mind Body Med—Advances in Mind Body Medicine

Adv Nutr—Advances in Nutrition

Adv Prev Med—Advances in Preventative Medicine

Alzheimers Dement—Alzheimer's & Dementia

Am Fam Physician—American Family Physician

Am J Cardiol—The American Journal of Cardiology

Am J Clin Nutr—The American Journal of Clinical Nutrition

Am J Epidemiol—American Journal of Epidemiology

Am J Geriatr Psychiatry—The American Journal of Geriatric Psychiatry

Am J Prev Med—American Journal of Preventative Medicine

Am J Psychiatry—American Journal of Psychiatry

Anc Sci—Ancient Science of Life

Ann Gen Psychiatry—Annals of General Psychiatry

Ann Neurol—Annals of Neurology

Ann Nutr Metab—Annals of Nutrition and Metabolism

Annu Rev Psychol—Annual Review of Psychology

Antiinflamm Antiallergy Agents Med Chem—Anti-inflammatory & Anti-Allergy Agents in Medicinal Chemistry

Arch Environ Health—Archives of Environmental Health

Auton Neurosci—Autonomic Neuroscience: Basic and Clinical

Biomol Ther—Biomolecules & Therapeutics

BMC Complement Altern Med—BMC Complementary and Alternative Medicine

BMJ—The BMJ

Brain Behav Immun—Brain, Behavior and Immunity

Brain Connect—Brain Connectivity

Brain Imaging Behav—Brain Imaging and Behavior

Brain Plast—Brain Plasticity

Br J Nutr—British Journal of Nutrition

Br J Pharmacol—British Journal of Pharmacology

Br J Psychol—British Journal of Psychology

Br J Sports Med—British Journal of Sports Medicine

Cereb Cortex—Cerebral Cortex

Chin Med—Chinese Medicine

Clin EEG Neurosci—Clinical EEG and Neuroscience

Clin Nutr—Clinical Nutrition

Clin Pract—Clinical Practice

Cochrane Database Syst Rev—Cochrane Database Systematic Reviews

Cureus—The Cureus Journal of Medical Science

Dev Cogn Neurosci—Developmental Cognitive Neuroscience

Environ Health Insights—Environmental Health Insights

Environ Health Perspect—Environmental Health Perspectives

Environ Sci Technol—Environmental Science & Technology

Eur J Social Psychology—European Journal of Social Psychology

Evid Based Complement Alternat Med—Evidence-Based Complementary and Alternative Medicine

FASEB—Federation of American Societies for Experimental Biology

Food Chem Toxicol—Food and Chemical Toxicology

Food Funct—Food & Function

Front Aging Neurosci—Frontiers in Aging Neuroscience

Front Hum Neurosci—Frontiers in Human Neuroscience

Front Integr Neurosci—Frontiers in Integrative Neuroscience

Front Neuroendocrinol—Frontiers in Neuroendocrinology

Front Nutr—Frontiers in Nutrition

Front Pharmacol—Frontiers in Pharmacology

Front Psychol—Frontiers in Psychology

Front Public Health—Frontiers in Public Health

Gen Hosp Psychiatry—General Hospital Psychiatry

Genes Nutr—Genes & Nutrition

Hum Brain Mapp—Human Brain Mapping

Integr Med (Encinitas)—Integrative Medicine

Int J Addict—The International Journal of the Addictions

Int J Alzheimers Dis—International Journal of Alzheimer's Disease

Int J Biochem Cell Biol—International Journal of Biochemistry & Cell Biology

Int J Exerc Sci—International Journal of Exercise Science

Int J Geriatr Psychiatry—International Journal of Geriatric Psychiatry

Int Psychogeriatr—International Psychogeriatrics

J Aging Health—Journal of Aging and Health

J Agric Food Chem—Journal of Agricultural and Food Chemistry

J Altern Complement Med—The Journal of Alternative and Complimentary Medicine

J Alzheimers Dis—Journal of Alzheimer's Disease

JAMA Neurol—JAMA Neurology

JAMA Otolaryngol Head Neck Surg—JAMA Otolaryngology—Head & Neck Surgery

J Am Coll Nutr—Journal of the American College of Nutrition

J Am Osteopath Assoc—The Journal of the American Osteopathic Association

J Ayurveda Integr Med—Journal of Ayurveda and Integrative Medicine

J Clin Diagn Res—Journal of Clinical and Diagnostic Research

J Clin Endocrinol Metab—Journal of Clinical Endocrinology and Metabolism

J Clin Invest—Journal of Clinical Investigation

J Comp Neurol—Journal of Comparative Neurology

J Exp Soc Psychol—Journal of Experimental Social Psychology

J Health Psychol—Journal of Health Psychology

J Health Soc Behav—Journal of Health and Social Behavior

J Hum Nutr Diet—Journal of Human Nutrition and Dietetics

J Inorg Biochem—Journal of Inorganic Biochemistry

J Neurosci—The Journal of Neuroscience

J Nurs Scholarsh—Journal of Nursing Scholarship

J Nutr—The Journal of Nutrition

J Nutr Health Aging—The Journal of Health, Nutrition and Aging

J Pain—The Journal of Pain

J Pers Soc Psychol—Journal of Personality and Social Psychology

J Physiol—The Journal of Physiology

J Psychoactive Drugs—Journal of Psychoactive Drugs

J Subst Abuse—Journal of Substance Abuse Treatment

Lancet Neurol—Lancet Neurology

Magnes Res—Magnesium Research

Med Gas Res—Medical Gas Research
Med Sci Sports Exerc—Medicine & Science in Sports & Exercise
Mol Psychiatry—Molecular Psychiatry
Mult Scler—Multiple Sclerosis Journal
Nat Commun—Nature Communications
Nat Hum Behav—Nature Human Behavior
Nat Med—Nature Medicine
Neurobiol Dis—Neurobiology of Disease
Neurobiol Learn Mem—Neurobiology of Learning and Memory
Neurochem Int—Neurochemistry International
Neurochem Res—Neurochemical Research
Neuropsychology—Neuropsychology
Neuropsychol Rev—Neuropsychology Review
NPJ Sci Learn—NPJ Science of Learning
Nutr Cancer—Nutrition and Cancer
Nutr J—Nutrition Journal
Nutr Neurosci—Nutritional Neuroscience
Obes Facts—Obesity Facts
Perspect Psychol Sci—Perspectives on Psychological Science
PLoS Genet—PLoS Genetics
PLoS Med—PLoS Medicine
Proc Natl Acad Sci USA—Proceedings of the National Academy of Sciences of the United States of America
Prog Neuropsychopharmacol Biol Psychiatry—Progress in Neuro-Psychopharmacology & Biological Psychiatry
Psychiatry Res—Psychiatry Research
Psychol Bull—Psychological Bulletin
Psychol Sci—Psychological Science
Psychopharmacol Bull—Psychopharmacology Bulletin
Risk Manag Healthc Policy—Risk Management and Healthcare Policy
Sci Pharm—Scientia Pharmaceutica
Soc Cogn Affect Neurosci—Social Cognitive and Affective Neuroscience
Soc Sci Med—Social Science & Medicine
Transl Psychiatry—Translational Psychiatry
Trends Neurosci Educ—Trends in Neuroscience and Education
West Indian Med J—West Indian Medical Journal

Chapter 1: Yes, You Can Change Your Brain

1. Bartucca J. The Most Complicated Object in the Universe. University of Connecticut. https://today.uconn.edu/2018/03/complicated-object-universe/. Published 2018.

2. Mayo Foundation for Medical Education and Research (MFMER). Stress Basics. https://www.mayoclinic.org/healthy-lifestyle/stress-management/basics/stress-basics/hlv-20049495. Published 2017. Accessed March 31, 2017.

3. Chetty S, Friedman AR, Taravosh-Lahn K, et al. Stress and Glucocorticoids Promote Oligodendrogenesis in the Adult Hippocampus. *Mol Psychiatry*. 2014;19(12):1275–83.

4. Thomson EM. Air Pollution, Stress, and Allostatic Load: Linking Systemic and Central Nervous System Impacts. *J Alzheimers Dis.* 2019;69(3):597–614.

5. National Institute of Environmental Health Sciences. Electric & Magnetic Fields. https://www.niehs.nih.gov/health/topics/agents/emf/index.cfm. Published 2018.

6. Kim JH, Lee JK, Kim HG, Kim KB, Kim HR. Possible Effects of Radiofrequency Electromagnetic Field Exposure on Central Nerve System. *Biomol Ther (Seoul)*. 2019;27(3):265–75.

7. Kim JH, Lee JK, Kim HG, Kim KB, Kim HR. Possible Effects of Radiofrequency Electromagnetic Field Exposure on Central Nerve System. *Biomol Ther (Seoul)*. 2019;27(3):265–75.

8. Bast T, Pezze M, McGarrity S. Cognitive deficits caused by prefrontal cortical and hippocampal neural disinhibition. *Br J Pharmacol.* 2017;174(19):3211–25.

9. Augusta Health. What Happens to Your Brain as You Age? https://www.augustahealth.com/health-focused/what-happens-to-your-brain-as-you-age. Published 2018.

10. Hartshorne JK, Germine LT. When Does Cognitive Functioning Peak? The Asynchronous Rise and Fall of Different Cognitive Abilities Across the Life Span. *Psychol Sci.* 2015;26(4):433–43.

11. Fortenbaugh FC, DeGutis J, Germine L, et al. Sustained Attention Across the Life Span in a Sample of 10,000: Dissociating Ability and Strategy. *Psychol Sci*. 2015;26(9):1497–1510.

12. Michel A. The Cognitive Upside of Aging. Association for Psychological Science. https://www.psychologicalscience.org/observer/the-cognitive-upside-of-aging. Published 2017. Accessed January 31, 2017.

13. Phillips M. The Mind at Midlife. American Psychological Association. https://www.apa.org/monitor/2011/04/mind-midlife. Published 2011. Accessed April 2011.

14. Michel A. The Cognitive Upside of Aging. Association for Psychological Science. https://www.psychologicalscience.org/observer/the-cognitive-upside-of-aging. Published 2017. Accessed January 31, 2017.

15. Taylor JL, Kennedy Q, Noda A, Yesavage JA. Pilot Age and Expertise Predict Flight Simulator Performance: A 3-Year Longitudinal Study. *Neurology*. 2007;68(9):648–54.

16. Blanchflower DG, Oswald AJ. Is Well-Being U-Shaped over the Life Cycle? *Soc Sci Med.* 2008;66(8):1733–49.

17. Williams LM, Brown KJ, Palmer D, et al. The Mellow Years?: Neural Basis of Improving Emotional Stability over Age. *J Neurosci.* 2006;26(24):6422–30.

18. Socci V, Tempesta D, Desideri G, De Gennaro L, Ferrara M. Enhancing Human Cognition with Cocoa Flavonoids. *Front Nutr.* 2017;4:19.

19. Brinol P, Petty RE, Wagner B. Body Posture Effects on Self-Evaluation: A Self-Validation Approach. *Eur J Social Psychology.* 2009;39(6):1053–64.

20. Sowndhararajan K, Kim S. Influence of Fragrances on Human Psychophysiological Activity: With Special Reference to Human Electroencephalographic Response. *Sci Pharm.* 2016;84(4):724–51.

Chapter 2: Brain Basics

1. Koch C. Does Brain Size Matter? *Scientific American Mind.* 2016(January–February):22–25.

2. Amen D. *Unleash the Power of the Female Brain: Supercharging Yours for Better Health, Energy, Mood, Focus and Sex.* New York: Crown, 2013.

3. Ingalhalikar M, Smith A, Parker D, et al. Sex Differences in the Structural Connectome of the Human Brain. *Proc Natl Acad Sci U S A.* 2014;111(2):823–28.

4. Rippon G. *Gender and Our Brains: How New Neuroscience Explodes the Myths of the Male and Female Minds.* New York: Pantheon, 2019.

5. Ross V. Numbers: The Nervous System, From 268-Mph Signals to Trillions of Synapses. *Discover Magazine.* http://www.discovermagazine.com/health/numbers-the-nervous-system-from-268-mph-signals-to-trillions-of-synapses. Published 2011.

6. Stanford University. What Is Your Reaction Time? http://virtuallabs.stanford.edu/tech/images/ReactionTime.SU-Tech.pdf. Published 2007.

7. Stanford University. What Is Your Reaction Time? http://virtuallabs.stanford.edu/tech/images/ReactionTime.SU-Tech.pdf. Published 2007.

8. Stone M. Could You Charge an iPhone with the Electricity in Your Brain? Gizmodo. https://gizmodo.com/could-you-charge-an-iphone-with-the-electricity-in-your-1722569935. Published 2015.

9. Clinical Neurology Specialists. What Is the Memory Capacity of a Human Brain? https://www.cnsnevada.com/what-is-the-memory-capacity-of-a-human-brain/.

10. Reber P. What Is the Memory Capacity of the Human Brain? *Scientific American* 2010.

11. Valentine RC, Valentine DL. *Neurons and the DHA Principle.* Boca Raton, Fla.: CRC Press / Taylor & Francis Group, 2019.

12. Herculano-Houzel S. The Human Brain in Numbers: A Linearly Scaled-Up Primate Brain. *Front Hum Neurosci.* 2009;3:31.

13. Burgess L. Left Brain vs. Right Brain: Fact and Fiction. Medical News Today. https://www.medicalnewstoday.com/articles/321037. Published 2018.

14. Burgess L. Left Brain vs. Right Brain: Fact and Fiction. Medical News Today. https://www.medicalnewstoday.com/articles/321037. Published 2018.

15. Reeves AG, Swenson RS. *Disorders of the Nervous System: A Primer.* Online version published by Dartmouth Medical School. https://www.dartmouth.edu/~dons/part_1/chapter_2.html. Published 2008.

16. Uylings HB, Jacobsen AM, Zilles K, Amunts K. Left-Right Asymmetry in Volume and Number of Neurons in Adult Broca's Area. *Cortex.* 2006;42(4):652–58.

17. Burgess L. Left Brain vs. Right Brain: Fact and Fiction. Medical News Today. https://www.medicalnewstoday.com/articles/321037. Published 2018.

18. Lemon RN, Edgley SA. Life Without a Cerebellum. *Brain.* 2010;133(3): 652–54.

19. Hamilton DM. Calming Your Brain During Conflict. *Harvard Business Review.* https://hbr.org/2015/12/calming-your-brain-during-conflict. Published 2015.

20. Schultz DH, Balderston NL, Baskin-Sommers AR, Larson CL, Helmstetter FJ. Corrigendum: Psychopaths Show Enhanced Amygdala Activation During Fear Conditioning. *Front Psychol.* 2017;8:1457.

21. Sohn E. Decoding the Neuroscience of Consciousness. *Nature.* https://www.nature.com/articles/d41586-019-02207-1. Published 2019.

22. Owen AM, Coleman MR, Boly M, Davis MH, Laureys S, Pickard JD. Detecting Awareness in the Vegetative State. *Science.* 2006;313(5792):1402.

23. Freud's Model of the Human Mind. Journal Psyche. http://journalpsyche.org/understanding-the-human-mind/.

24. Freud's Model of the Human Mind. Journal Psyche. http://journalpsyche.org/understanding-the-human-mind/.

25. Goriounova NA, Mansvelder HD. Genes, Cells and Brain Areas of Intelligence. *Front Hum Neurosci.* 2019;13:44.

26. Goriounova NA, Mansvelder HD. Genes, Cells and Brain Areas of Intelligence. *Front Hum Neurosci.* 2019;13:44.

Thomas MS. Do More Intelligent Brains Retain Heightened Plasticity for Longer in Development? A Computational Investigation. *Dev Cogn Neurosci.* 2016;19:258–69.

27. Stevens AP. Learning Rewires the Brain. Science News for Students. https://www.sciencenewsforstudents.org/article/learning-rewires-brain. Published 2014.

28. Small GW, Silverman DH, Siddarth P, et al. Effects of a 14-Day Healthy Longevity Lifestyle Program on Cognition and Brain Function. *Am J Geriatr Psychiatry*. 2006;14(6):538–45.

29. American Psychological Association. Believing You Can Get Smarter Makes You Smarter. Published 2003.

Aronson J, Fried CB, Good C. Reducing the Effects of Stereotype Threat on African American College Students by Shaping Theories of Intelligence. *J Exp Soc Psychol*.2002;38(2):113–25.

30. Shenk D. The Truth About IQ. *The Atlantic*. https://www.theatlantic.com/national/archive/2009/07/the-truth-about-iq/22260/. Published 2009.

National Academies of Sciences and Medicine; Division of Behavioral and Social Sciences and Education; Board on Children, Youth, and Families; Committee on Supporting the Parents of Young Children. *Parenting Matters: Supporting Parents of Children Age 0–8*. Washington, D.C.: National Academies Press, 2016.

31. Whale Facts. Sperm Whale Brain and Intelligence. https://www.whalefacts.org/sperm-whale-brain/.

32. WebMD. How Your Brain Works: Myths and Facts. https://www.webmd.com/brain/rm-quiz-brain-works.

33. Muench K. Pain in the Brain. NeuWrite West. http://www.neuwritewest.org/blog/pain-in-the-brain. Published 2015.

34. Wake Forest Baptist Medical Center. Neuroscientists Explain How the Sensation of Brain Freeze Works. Science Daily. https://www.sciencedaily.com/releases/2013/05/130522095335.htm. Published 2013.

35. Nordqvist J. Why Does Ice Cream Cause Brain Freeze? Medical News Today. https://www.medicalnewstoday.com/articles/244458. Published 2017.

36. Richards BA, Frankland PW. The Persistence and Transience of Memory. *Neuron*. 2017;94(6):1071–84.

37. Riccelli R, Toschi N, Nigro S, Terracciano A, Passamonti L. Surface-Based Morphometry Reveals the Neuroanatomical Basis of the Five-Factor Model of Personality. *Soc Cogn Affect Neurosci*. 2017;12(4):671–84.

38. Riccelli R, Toschi N, Nigro S, Terracciano A, Passamonti L. Surface-Based Morphometry Reveals the Neuroanatomical Basis of the Five-Factor Model of Personality. *Soc Cogn Affect Neurosci*. 2017;12(4):671–84.

39. Alzheimer's Association. Alzheimer's and Dementia: Facts and Figures. https://www.alz.org/alzheimers-dementia/facts-figures.

40. TraumaticBrainInjury.com. Mild TBI Symptoms. https://www.traumaticbraininjury.com/mild-tbi-symptoms/. Published 2019.

41. Centers for Disease Control and Prevention. CDC Announces Updated Information to Help Physicians Recognize and Manage Concussions Early. https://www.cdc.gov/media/pressrel/2007/r070607.htm. Published 2007.

42. National Institute of Mental Health. Major Depression. https://www.nimh.nih.gov/health/statistics/major-depression.shtml. Published 2019.

43. Brody DJ, Pratt LA, Hughes JP. Prevalence of Depression Among Adults Aged 20 and Over: United States, 2013–2016. NCHS Data Brief, no 303. National Center for Health Statistics. Centers for Disease Control and Prevention. https://www.cdc.gov/nchs/products/databriefs/db303.htm. Published 2018.

44. Benjamin EJ, Blaha MJ, Chiuve SE, et al. Heart Disease and Stroke Statistics—2017 Update: A Report from the American Heart Association. *Circulation*. 2017;135(10):e146–e603.

Chapter 3: The Better Brain Diet

1. Martinez Steele E, Popkin BM, Swinburn B, Monteiro CA. The Share of Ultra-Processed Foods and the Overall Nutritional Quality of Diets in the US: Evidence from a Nationally Representative Cross-Sectional Study. *Population Health Metrics* 2017;15(1):6.

2. Office of Disease Prevention and Health Promotion. 2015–2020 Dietary Guidelines for Americans—Cut Down on Added Sugars. https://health.gov/sites/default/files/2019-10/DGA_Cut-Down-On-Added-Sugars.pdf. Published 2016.

3. Srour B, Fezeu LK, Kesse-Guyot E, et al. Ultra-Processed Food Intake and Risk of Cardiovascular Disease: Prospective Cohort Study (NutriNet-Sante). *BMJ*. 2019;365:l1451.

Rico-Campa A, Martinez-Gonzalez MA, Alvarez-Alvarez I, et al. Association Between Consumption of Ultra-Processed Foods and All Cause Mortality: SUN Prospective Cohort Study. *BMJ*. 2019;365:l1949.

4. Chang CY, Ke DS, Chen JY. Essential Fatty Acids and Human Brain. *Acta Neurol Taiwan*. 2009;18(4):231–41.

5. National Institutes of Health. Office of Dietary Supplements. Omega-3 Fatty Acids. https://ods.od.nih.gov/factsheets/Omega3FattyAcids-HealthProfessional/. Published 2019.

6. Lloyd-Jones DM, Hong Y, Labarthe D, et al. Defining and Setting National Goals for Cardiovascular Health Promotion and Disease Reduction: the American Heart Association's Strategic Impact Goal Through 2020 and Beyond. *Circulation*. 2010;121(4):586–613.

7. Chang CY, Ke DS, Chen JY. Essential Fatty Acids and Human Brain. *Acta Neurol Taiwan*. 2009;18(4):231–41.

Papanikolaou Y, Brooks J, Reider C, Fulgoni VL, 3rd. U.S. adults are not meeting recommended levels for fish and omega-3 fatty acid intake: results of an analysis using observational data from NHANES 2003–2008. *Nutr J*. 2014;13:31.

8. National Institutes of Health. Office of Dietary Supplements. Omega-3 Fatty Acids. https://ods.od.nih.gov/factsheets/Omega3FattyAcids-HealthProfessional/. Published 2019.

9. Okereke OI, Rosner BA, Kim DH, et al. Dietary Fat Types and 4-Year Cognitive Change in Community-Dwelling Older Women. *Ann Neurol.* 2012;72(1):124–34.

10. Dean W, English J. Medium Chain Triglycerides (MCTs): Beneficial Effects on Energy, Atherosclerosis and Aging. Nutrition Review. https://nutritionreview.org/2013/04/medium-chain-triglycerides-mcts/. Published 2013.

11. Dean W, English J. Medium Chain Triglycerides (MCTs): Beneficial Effects on Energy Atherosclerosis and Aging. Nutrition Review. https://nutritionreview.org/2013/04/medium-chain-triglycerides-mcts/. Published 2013.

12. Swaminathan A, Jicha GA. Nutrition and Prevention of Alzheimer's Dementia. *Front Aging Neurosci.* 2014;6:282.

Croteau E, Castellano CA, Richard MA, et al. Ketogenic Medium Chain Triglycerides Increase Brain Energy Metabolism in Alzheimer's Disease. *J Alzheimers Dis.* 2018;64(2):551–61.

13. Wengreen H, Munger RG, Cutler A, et al. Prospective Study of Dietary Approaches to Stop Hypertension- and Mediterranean-Style Dietary Patterns and Age-Related Cognitive Change: The Cache County Study on Memory, Health and Aging. *Am J Clin Nutr.* 2013;98(5):1263–71.

14. Ozawa M, Shipley M, Kivimaki M, Singh-Manoux A, Brunner EJ. Dietary Pattern, Inflammation and Cognitive Decline: The Whitehall II Prospective Cohort Study. *Clin Nutr.* 2017;36(2):506–12.

15. Burgess L. 12 Foods to Boost Brain Function. Medical News Today. https://www.medicalnewstoday.com/articles/324044. Published 2020.

16. Hwang SL, Shih PH, Yen GC. Neuroprotective Effects of Citrus Flavonoids. *J Agric Food Chem.* 2012;60(4):877–85.

17. Burgess L. 12 Foods to Boost Brain Function. Medical News Today. https://www.medicalnewstoday.com/articles/324044. Published 2020.

18. Burgess L. 12 Foods to Boost Brain Function. Medical News Today. https://www.medicalnewstoday.com/articles/324044. Published 2020.

19. Berk L, Lohman E, Bains G, et al. Nuts and Brain Health: Nuts Increase EEG Power Spectral Density (μV&[sup2]) for Delta Frequency (1–3Hz) and Gamma Frequency (31–40 Hz) Associated with Deep Meditation, Empathy, Healing, as well as Neural Synchronization, Enhanced Cognitive Processing, Recall, and Memory All Beneficial For Brain Health. *FASEB,* 2017.

20. Poulose SM, Miller MG, Shukitt-Hale B. Role of Walnuts in Maintaining Brain Health with Age. *J Nutr.* 2014;144(4 Suppl):561S–66S.

21. Medawar E, Huhn S, Villringer A, Veronica Witte A. The Effects of Plant-Based Diets on the Body and the Brain: A Systematic Review. *Transl Psychiatry.* 2019;9(1):226.

22. Medawar E, Huhn S, Villringer A, Veronica Witte A. The Effects of Plant-Based Diets on the Body and the Brain: A Systematic Review. *Transl Psychiatry*. 2019;9(1):226.

23. De la Monte SM, Tong M. Mechanisms of Nitrosamine-Mediated Neurodegeneration: Potential Relevance to Sporadic Alzheimer's Disease. *J Alzheimers Dis*. 2009;17(4):817–25.

24. Ward RJ, Zucca FA, Duyn JH, Crichton RR, Zecca L. The Role of Iron in Brain Ageing and Neurodegenerative Disorders. *Lancet Neurol*. 2014;13(10):1045–60.

25. Romeu M, Aranda N, Giralt M, Ribot B, Nogues MR, Arija V. Diet, Iron Biomarkers and Oxidative Stress in a Representative Sample of Mediterranean Population. *Nutr J*. 2013;12:102.

26. Freeman LR, Haley-Zitlin V, Rosenberger DS, Granholm AC. Damaging Effects of a High-Fat Diet to the Brain and Cognition: A Review of Proposed Mechanisms. *Nutr Neurosci*. 2014;17(6):241–51.

27. Getaneh G, Mebrat A, Wubie A, Kendie H. Review on Goat Milk Composition and Its Nutritive Value. *Journal of Nutrition and Health Sciences*. 2016;3(4):1–10.

28. Medawar E, Huhn S, Villringer A, Veronica Witte A. The Effects of Plant-Based Diets on the Body and the Brain: A Systematic Review. *Transl Psychiatry*. 2019;9(1):226.

29. Harvard T.H. Chan School of Public Health. Straight Talk About Soy. https://www.hsph.harvard.edu/nutritionsource/soy/.

30. Oldways Whole Grains Council. Whole Grain Protein Power! https://wholegrainscouncil.org/blog/2014/02/whole-grain-protein-power. Published 2014.

31. Mayer EA, Tillisch K, Gupta A. Gut/Brain Axis and the Microbiota. *J Clin Invest*. 2015;125(3):926–38.

Clapp M, Aurora N, Herrera L, Bhatia M, Wilen E, Wakefield S. Gut Microbiota's Effect on Mental Health: The Gut-Brain Axis. *Clin Pract*. 2017;7(4):987.

32. Medawar E, Huhn S, Villringer A, Veronica Witte A. The Effects of Plant-Based Diets on the Body and the Brain: A Systematic Review. *Transl Psychiatry*. 2019;9(1):226.

33. Moore J, Fung J. *The Complete Guide to Fasting: Heal Your Body Through Intermittent, Alternate-Day, and Extended Fasting*. Las Vegas, Nev.: Victory Belt Publishing, 2016.

Anton SD, Moehl K, Donahoo WT, et al. Flipping the Metabolic Switch: Understanding and Applying the Health Benefits of Fasting. *Obesity (Silver Spring)*. 2018;26(2):254–68.

34. Li L, Wang Z, Zuo Z. Chronic Intermittent Fasting Improves Cognitive Functions and Brain Structures in Mice. *PLoS One*. 2013;8(6):e66069.

35. Morris MC, Tangney CC, Wang Y, Sacks FM, Bennett DA, Aggarwal NT. MIND Diet Associated with Reduced Incidence of Alzheimer's Disease. *Alzheimers Dement*. 2015;11(9):1007–14.

Chapter 4: The Better Brain Workout

1. Zhang R, Parker R, Zhu YS, et al. Aerobic Exercise Training Increases Brain Perfusion in Elderly Women. *FASEB*. 2011;25(1 Suppl).

2. Alfini AJ, Weiss LR, Leitner BP, Smith TJ, Hagberg JM, Smith JC. Hippocampal and Cerebral Blood Flow After Exercise Cessation in Master Athletes. *Front Aging Neurosci*. 2016;8:184.

3. Cohen DL, Wintering N, Tolles V, et al. Cerebral Blood Flow Effects of Yoga Training: Preliminary Evaluation of 4 Cases. *J Altern Complement Med*. 2009;15(1):9–14.

4. Experimental Biology. How walking benefits the brain: Researchers Show That Foot's Impact Helps Control, Increase the Amount of Blood Sent to the Brain. Science Daily. https://www.sciencedaily.com/releases/2017/04/170424141340.htm. Published 2017.

5. Eriksson PS, Perfilieva E, Bjork-Eriksson T, et al. Neurogenesis in the Adult Human Hippocampus. *Nat Med*. 1998;4(11):1313–17.

6. Van Praag H, Christie BR, Sejnowski TJ, Gage FH. Running Enhances Neurogenesis, Learning, and Long-Term Potentiation in Mice. *Proc Natl Acad Sci U.S.A*. 1999;96(23):13427–31.

7. Nokia MS, Lensu S, Ahtiainen JP, et al. Physical Exercise Increases Adult Hippocampal Neurogenesis in Male Rats Provided It Is Aerobic and Sustained. *J Physiol*. 2016;594(7):1855–73.

Harvard Health Publishing. Can You Grow New Brain Cells? https://www.health.harvard.edu/mind-and-mood/can-you-grow-new-brain-cells. Published 2016.

8. Leiter O, Seidemann S, Overall RW, et al. Exercise-Induced Activated Platelets Increase Adult Hippocampal Precursor Proliferation and Promote Neuronal Differentiation. *Stem Cell Reports*. 2019;12(4):667–79.

9. Nokia MS, Lensu S, Ahtiainen JP, et al. Physical Exercise Increases Adult Hippocampal Neurogenesis in Male Rats Provided It Is Aerobic and Sustained. *J Physiol*. 2016;594(7):1855–73.

10. Hoang TD, Reis J, Zhu N, et al. Effect of Early Adult Patterns of Physical Activity and Television Viewing on Midlife Cognitive Function. *JAMA Psychiatry*. 2016;73(1):73–79.

11. Firth J, Stubbs B, Vancampfort D, et al. Effect of Aerobic Exercise on Hippocampal Volume in Humans: A Systematic Review and Meta-analysis. *Neuroimage*. 2018;166:230–38.

12. Rush University Medical Center. Everyday Activities Associated with More Gray Matter in Brains of Older Adults: Study Measured Amount of Life-

style Physical Activity Such as House Work, Dog Walking and Gardening. Science Daily. https://www.sciencedaily.com/releases/2018/02/180214093828.htm. Published 2018.

13. Burzynska AZ, Chaddock-Heyman L, Voss MW, et al. Physical Activity and Cardiorespiratory Fitness Are Beneficial for White Matter in Low-Fit Older Adults. *PLoS One*. 2014;9(9):e107413.

14. Gothe NP, Khan I, Hayes J, Erlenbach E, Damoiseaux JS. Yoga Effects on Brain Health: A Systematic Review of the Current Literature. *Brain Plast*. 2019;5(1):105–22.

15. Godman H. Regular Exercise Changes the Brain to Improve Memory, Thinking Skills. Harvard Health Publishing. https://www.health.harvard.edu/blog/regular-exercise-changes-brain-improve-memory-thinking-skills-201404097110. Published 2018.

16. Raichlen DA, Bharadwaj PK, Fitzhugh MC, et al. Differences in Resting State Functional Connectivity Between Young Adult Endurance Athletes and Healthy Controls. *Front Hum Neurosci*. 2016;10:610.

17. Chen C, Nakagawa S, An Y, Ito K, Kitaichi Y, Kusumi I. The Exercise-Glucocorticoid Paradox: How Exercise Is Beneficial to Cognition, Mood, and the Brain While Increasing Glucocorticoid Levels. *Front Neuroendocrinol*. 2017;44:83–102.

18. Greenwood BN, Kennedy S, Smith TP, Campeau S, Day HE, Fleshner M. Voluntary Freewheel Running Selectively Modulates Catecholamine Content in Peripheral Tissue and c-Fos Expression in the Central Sympathetic Circuit Following Exposure to Uncontrollable Stress in Rats. *Neuroscience*. 2003;120(1):269–81.

19. Mischel NA, Llewellyn-Smith IJ, Mueller PJ. Physical (In)Activity-Dependent Structural Plasticity in Bulbospinal Catecholaminergic Neurons of Rat Rostral Ventrolateral Medulla. *J Comp Neurol*. 2014;522(3):499–513.

20. Yorks DM, Frothingham CA, Schuenke MD. Effects of Group Fitness Classes on Stress and Quality of Life of Medical Students. *J Am Osteopath Assoc*. 2017;117(11):e17–e25.

21. Van Den Berg AE, Custers MH. Gardening Promotes Neuroendocrine and Affective Restoration from Stress. *J Health Psychol*. 2011;16(1):3–11.

22. Harvard Health Publishing. Exercise Is an All-Natural Treatment to Fight Depression. https://www.health.harvard.edu/mind-and-mood/exercise-is-an-all-natural-treatment-to-fight-depression. Published 2013.

Blumenthal JA, Smith PJ, Hoffman BM. Is Exercise a Viable Treatment for Depression? *ACSMs Health Fit J*. 2012;16(4):14–21.

23. Castrén E, Kojima M. Brain-Derived Neurotrophic Factor in Mood Disorders and Antidepressant Treatments. *Neurobiol Dis*. 2017;97(Pt B):119–26.

24. Weir K. The Exercise Effect. American Psychological Association. https://www.apa.org/monitor/2011/12/exercise. Published 2011.

25. Weir K. The Exercise Effect. American Psychological Association. https://www.apa.org/monitor/2011/12/exercise. Published 2011.

26. Barton J, Pretty J. What Is the Best Dose of Nature and Green Exercise for Improving Mental Health? A Multi-study Analysis. *Environ Sci Technol.* 2010;44(10):3947–55.

27. Bratman GN, Hamilton JP, Hahn KS, Daily GC, Gross JJ. Nature Experience Reduces Rumination and Subgenual Prefrontal Cortex Activation. *Proc Natl Acad Sci U.S.A.* 2015;112(28):8567–72.

28. Dolezal BA, Neufeld EV, Boland DM, Martin JL, Cooper CB. Interrelationship Between Sleep and Exercise: A Systematic Review. *Adv Prev Med.* 2017;2017:1364387.

29. National Sleep Foundation. How Exercise Affects Sleep. Sleep.org. https://www.sleep.org/articles/exercise-affects-sleep/. Published 2020.

30. Bankar MA, Chaudhari SK, Chaudhari KD. Impact of Long Term Yoga Practice on Sleep Quality and Quality of Life in the Elderly. *J Ayurveda Integr Med.* 2013;4(1):28–32.

31. Johns Hopkins Medicine. Exercising for Better Sleep. https://www.hopkinsmedicine.org/health/wellness-and-prevention/exercising-for-better-sleep.

32. Mead MN. Benefits of Sunlight: A Bright Spot for Human Health. *Environ Health Perspect.* 2008;116(4):A160–A167.

33. Erion JR, Wosiski-Kuhn M, Dey A, et al. Obesity Elicits Interleukin 1-Mediated Deficits in Hippocampal Synaptic Plasticity. *J Neurosci.* 2014;34(7):2618–31.

Rhea EM, Salameh TS, Logsdon AF, Hanson AJ, Erickson MA, Banks WA. Blood-Brain Barriers in Obesity. *AAPS J.* 2017;19(4):921–30.

34. Rhea EM, Salameh TS, Logsdon AF, Hanson AJ, Erickson MA, Banks WA. Blood-Brain Barriers in Obesity. *AAPS J.* 2017;19(4):921–30.

35. Willeumier KC, Taylor DV, Amen DG. Elevated BMI is Associated with Decreased Blood Flow in the Prefrontal Cortex Using SPECT Imaging in Healthy Adults. *Obesity (Silver Spring).* 2011;19(5):1095–97.

36. Willeumier K, Taylor DV, Amen DG. Elevated Body Mass in National Football League Players Linked to Cognitive Impairment and Decreased Prefrontal Cortex and Temporal Pole Activity. *Transl Psychiatry.* 2012;2(1):e68.

37. Erion JR, Wosiski-Kuhn M, Dey A, et al. Obesity Elicits Interleukin 1-Mediated Deficits in Hippocampal Synaptic Plasticity. *J Neurosci.* 2014;34(7):2618–31.

38. Kullmann S, Wagner L, Veit R, et al. Exercise Improves Brain Insulin Action and Executive Function in Adults with Overweight and Obesity. Paper presented at: Society for the Study of Ingestive Behavior 27th Annual Meeting, 2019; Utrecht, Netherlands.

39. Charvat M. Why Exercise Is Good for Your Brain. *Psychology Today.* https://www.psychologytoday.com/us/blog/the-fifth-vital-sign/201901/why-exercise-is-good-your-brain. Published 2019.

40. Lin WY, Chan CC, Liu YL, Yang AC, Tsai SJ, Kuo PH. Performing Different Kinds of Physical Exercise Differentially Attenuates the Genetic Effects on Obesity Measures: Evidence from 18,424 Taiwan Biobank Participants. *PLoS Genet.* 2019;15(8):e1008277.

41. Viana RB, Naves JPA, Coswig VS, et al. Is Interval Training the Magic Bullet for Fat Loss? A Systematic Review and Meta-analysis Comparing Moderate-Intensity Continuous Training with High-Intensity Interval Training (HIIT). *Br J Sports Med.* 2019;53(10):655–64.

42. Shah C, Beall EB, Frankemolle AM, et.al. Exercise Therapy for Parkinson's Disease: Pedaling Rate Is Related to Changes in Motor Connectivity. *Brain Connect.* 2016; 6(1):25–36.

43. Tarumi T, Rossetti H, Thomas BP, et al. Exercise Training in Amnestic Mild Cognitive Impairment: A One-Year Randomized Controlled Trial. *J Alzheimers Dis.* 2019;71(2):421–33.

Chapter 5: The Supplement Offensive

1. Amen DG, Wu JC, Taylor D, Willeumier K. Reversing Brain Damage in Former NFL Players: Implications for Traumatic Brain Injury and Substance Abuse Rehabilitation. *J Psychoactive Drugs.* 2011;43(1):1–5.

2. Amen DG, Taylor DV, Ojala K, Kaur J, Willeumier K. Effects of Brain-Directed Nutrients on Cerebral Blood Flow and Neuropsychological Testing: A Randomized, Double-Blind, Placebo-Controlled, Crossover Trial. *Adv Mind Body Med.* 2013;27(2):24–33.

3. *The Power of Seafood 2019: An In-Depth Look at Seafood Through the Shoppers' Eyes.* Arlington, Va.: Food Marketing Institute, 2019.

4. Lee HK, Kim SY, Sok SR. Effects of Multivitamin Supplements on Cognitive Function, Serum Homocysteine Level, and Depression of Korean Older Adults with Mild Cognitive Impairment in Care Facilities. *J Nurs Scholarsh.* 2016;48(3):223–31.

5. Fulgoni VL 3rd, Keast DR, Bailey RL, Dwyer J. Foods, Fortificants, and Supplements: Where Do Americans Get Their Nutrients? *J Nutr.* 2011;141(10):1847–54.

Drake VJ. Micronutrient Inadequacies in the US Population: An Overview. Linus Pauling Institute. Oregon State University. Published 2017.

6. Akbari E, Asemi Z, Daneshvar Kakhaki R, et al. Effect of Probiotic Supplementation on Cognitive Function and Metabolic Status in Alzheimer's Disease: A Randomized, Double-Blind and Controlled Trial. *Front Aging Neurosci.* 2016;8:256.

7. Anjum I, Jaffery SS, Fayyaz M, Samoo Z, Anjum S. The Role of Vitamin D in Brain Health: A Mini Literature Review. *Cureus.* 2018;10(7):e2960.

8. Banerjee A, Khemka VK, Ganguly A, Roy D, Ganguly U, Chakrabarti S. Vitamin D and Alzheimer's Disease: Neurocognition to Therapeutics. *Int J Alzheimers Dis*. 2015;2015:192747.

9. National Institutes of Health. Office of Dietary Supplements. Vitamin D Fact Sheet for Health Professionals. https://ods.od.nih.gov/factsheets/VitaminD-HealthProfessional/. Published 2019.

10. Nuttall JR, Oteiza PI. Zinc and the Aging Brain. *Genes Nutr*. 2014;9(1):379.

Prasad AS. Discovery of Human Zinc Deficiency: Its Impact on Human Health and Disease. *Adv Nutr*. 2013;4(2):176–90.

11. Solovyev ND. Importance of Selenium and Selenoprotein for Brain Function: From Antioxidant Protection to Neuronal Signalling. *J Inorg Biochem*. 2015;153:1–12.

12. Alizadeh M, Kheirouri S. Curcumin Reduces Malondialdehyde and Improves Antioxidants in Humans with Diseased Conditions: A Comprehensive Meta-analysis of Randomized Controlled Trials. *Biomedicine (Taipei)*. 2019;9(4):23.

13. Aggarwal BB, Harikumar KB. Potential Therapeutic Effects of Curcumin, the Anti-inflammatory Agent, Against Neurodegenerative, Cardiovascular, Pulmonary, Metabolic, Autoimmune and Neoplastic Diseases. *Int J Biochem Cell Biol*. 2009;41(1):40–59.

14. Wang R, Li YH, Xu Y, et al. Curcumin Produces Neuroprotective Effects via Activating Brain-Derived Neurotrophic Factor/TrkB-Dependent MAPK and PI-3K Cascades in Rodent Cortical Neurons. *Prog Neuropsychopharmacol Biol Psychiatry*. 2010;34(1):147–53.

15. Small GW, Siddarth P, Li Z, et al. Memory and Brain Amyloid and Tau Effects of a Bioavailable Form of Curcumin in Non-Demented Adults: A Double-Blind, Placebo-Controlled 18-Month Trial. *Am J Geriatr Psychiatry*. 2018;26(3):266–77.

16. Hewlings SJ, Kalman DS. Curcumin: A Review of Its Effects on Human Health. *Foods*. 2017;6(10).

17. Tayyem RF, Heath DD, Al-Delaimy WK, Rock CL. Curcumin Content of Turmeric and Curry Powders. *Nutr Cancer*. 2006;55(2):126–31.

18. Higdon J, Drake VJ, Delage B. Curcumin. Linus Pauling Institute. Oregon State University. https://lpi.oregonstate.edu/mic/dietary-factors/phytochemicals/curcumin. Published 2016.

19. Reynolds EH. Folic Acid, Ageing, Depression, and Dementia. *BMJ*. 2002;324(7352):1512–15.

20. Vogiatzoglou A, Refsum H, Johnston C, et al. Vitamin B_{12} Status and Rate of Brain Volume Loss in Community-Dwelling Elderly. *Neurology*. 2008;71(11):826–32.

Moore E, Mander A, Ames D, Carne R, Sanders K, Watters D. Cognitive Impairment and Vitamin B_{12}: A Review. *Int Psychogeriatr*. 2012;24(4):541–56.

21. Penninx BW, Guralnik JM, Ferrucci L, Fried LP, Allen RH, Stabler SP. Vitamin B(12) Deficiency and Depression in Physically Disabled Older Women: Epidemiologic Evidence from the Women's Health and Aging Study. *Am J Psychiatry*. 2000;157(5):715–21.

22. Moore E, Mander A, Ames D, Carne R, Sanders K, Watters D. Cognitive Impairment and Vitamin B_{12}: A Review. *Int Psychogeriatr.* 2012;24(4):541–56.

23. Paul C, Brady DM. Comparative Bioavailability and Utilization of Particular Forms of B_{12} Supplements With Potential to Mitigate B_{12}-Related Genetic Polymorphisms. *Integr Med (Encinitas)*. 2017;16(1):42–49.

24. Kim MK, Sasazuki S, Sasaki S, Okubo S, Hayashi M, Tsugane S. Effect of Five-Year Supplementation of Vitamin C on Serum Vitamin C Concentration and Consumption of Vegetables and Fruits in Middle-Aged Japanese: A Randomized Controlled Trial. *J Am Coll Nutr.* 2003;22(3):208–16.

25. Paleologos M, Cumming RG, Lazarus R. Cohort Study of Vitamin C Intake and Cognitive Impairment. *Am J Epidemiol.* 1998;148(1):45–50.

26. Michels A. Questions About Vitamin C. Linus Pauling Institute. Oregon State University. http://blogs.oregonstate.edu/linuspaulinginstitute/2015/05/28/questions-about-vitamin-c/. Published 2015.

27. Slutsky I, Abumaria N, Wu LJ, et al. Enhancement of Learning and Memory by Elevating Brain Magnesium. *Neuron*. 2010;65(2):165–77.

28. Hoane MR. The role of magnesium therapy in learning and memory. In: Vink R, Nechifor M, eds. *Magnesium in the Central Nervous System*. Adelaide, Australia: University of Adelaide Press, 2011.

29. Walker AF, Marakis G, Christie S, Byng M. Mg Citrate Found More Bioavailable Than Other Mg Preparations in a Randomised, Double-Blind Study. *Magnes Res*. 2003;16(3):183–91.

30. Monsef A, Shahidi S, Komaki A. Influence of Chronic Coenzyme Q10 Supplementation on Cognitive Function, Learning, and Memory in Healthy and Diabetic Middle-Aged Rats. *Neuropsychobiology*. 2019;77(2):92–100.

31. Stough C, Nankivell M, Camfield DA, et al. CoQ10 and Cognition: A Review and Study Protocol for a 90-Day Randomized Controlled Trial Investigating the Cognitive Effects of Ubiquinol in the Healthy Elderly. *Front Aging Neurosci*. 2019;11:103.

32. Ochiai A, Itagaki S, Kurokawa T, Kobayashi M, Hirano T, Iseki K. Improvement in Intestinal Coenzyme Q10 Absorption by Food Intake. *Yakugaku Zasshi*. 2007;127(8):1251–54.

33. Glade MJ, Smith K. Phosphatidylserine and the Human Brain. *Nutrition*. 2015;31(6):781–86.

34. Glade MJ, Smith K. Phosphatidylserine and the Human Brain. *Nutrition*. 2015;31(6):781–86.

35. Amaducci L. Phosphatidylserine in the Treatment of Alzheimer's Disease: Results of a Multicenter Study. *Psychopharmacol Bull*. 1988;24(1):130–34.

Crook T, Petrie W, Wells C, Massari DC. Effects of Phosphatidylserine in Alzheimer's Disease. *Psychopharmacol Bull.* 1992;28(1):61–66.

36. Benton D, Donohoe RT, Sillance B, Nabb S. The Influence of Phosphatidylserine Supplementation on Mood and Heart Rate When Faced with an Acute Stressor. *Nutr Neurosci.* 2001;4(3):169–78.

37. Hirayama S, Terasawa K, Rabeler R, et al. The Effect of Phosphatidylserine Administration on Memory and Symptoms of Attention-Deficit Hyperactivity Disorder: A Randomised, Double-Blind, Placebo-Controlled Clinical Trial. *J Hum Nutr Diet.* 2014;27 Suppl 2:284–91.

38. Purves D, Augustine GJ, Fitzpatrick D, et al. *Neuroscience. 2nd Edition.* Sunderland, Mass.: Sinauer Associates, 2001.

39. Wiklund O, Fager G, Andersson A, Lundstam U, Masson P, Hultberg B. N-acetylcysteine Treatment Lowers Plasma Homocysteine but Not Serum Lipoprotein(a) Levels. *Atherosclerosis.* 1996;119(1):99–106.

40. Lake J. Acetyl-l-carnitine: Important for Mental Health. Psychology Today. https://www.psychologytoday.com/us/blog/integrative-mental-health-care/201710/acetyl-l-carnitine-important-mental-health. Published 2017.

41. Smeland OB, Meisingset TW, Borges K, Sonnewald U. Chronic Acetyl-L-carnitine Alters Brain Energy Metabolism and Increases Noradrenaline and Serotonin Content in Healthy Mice. *Neurochem Int.* 2012;61(1):100–107.

42. Morgan AJ, Jorm AF. Self-Help Interventions for Depressive Disorders and Depressive Symptoms: A Systematic Review. *Ann Gen Psychiatry.* 2008;7:13.

43. Qian ZM, Ke Y. Huperzine A: Is It an Effective Disease-Modifying Drug for Alzheimer's Disease? *Front Aging Neurosci.* 2014;6:216.

44. *Chemical Information Review Document for Vinpocetine.* National Toxicology Program. National Institute of Environmental Health Sciences. 2013.

45. Valikovics A. Investigation of the Effect of Vinpocetine on Cerebral Blood Flow and Cognitive Functions. *Ideggyogy Sz.* 2007;60(7–8):301–10 (*Hungarian*).

46. Sierpina VS, Wollschlaeger B, Blumenthal M. Ginkgo Biloba. *Am Fam Physician.* 2003;68(5):923–26.

47. Sierpina VS, Wollschlaeger B, Blumenthal M. Ginkgo Biloba. *Am Fam Physician.* 2003;68(5):923–26.

Birks J, Grimley EV, Van Dongen M. Ginkgo biloba for cognitive impairment and dementia. *Cochrane Database Syst Rev.* 2002(4):CD003120.

48. Sierpina VS, Wollschlaeger B, Blumenthal M. Ginkgo Biloba. *Am Fam Physician.* 2003;68(5):923–26.

49. Molz P, Schröder N. Potential Therapeutic Effects of Lipoic Acid on Memory Deficits Related to Aging and Neurodegeneration. *Front Pharmacol.* 2017;8:849.

Chapter 6: The Hydration Offensive

1. Ericson J. 75% of Americans May Suffer from Chronic Dehydration, According to Doctors. Medical Daily. Published 2013.

2. Lieberman HR. Hydration and Cognition: A Critical Review and Recommendations for Future Research. *J Am Coll Nutr.* 2007;26(5 Suppl): 555S–561S.

3. Riebl SK, Davy BM. The Hydration Equation: Update on Water Balance and Cognitive Performance. *ACSMs Health Fit J.* 2013;17(6):21–28.

4. Wittbrodt MT, Millard-Stafford M. Dehydration Impairs Cognitive Performance: A Meta-analysis. *Med Sci Sports Exerc.* 2018;50(11):2360–68.

5. Pross N, Demazières A, Girard N, et al. Influence of Progressive Fluid Restriction on Mood and Physiological Markers of Dehydration in Women. *Br J Nutr.* 2013;109(2):313–21.

6. Kempton MJ, Ettinger U, Foster R, et al. Dehydration Affects Brain Structure and Function in Healthy Adolescents. *Hum Brain Mapp.* 2011;32(1):71–79.

7. Danone Nutricia Research. Hydration, Mood State and Cognitive Function. Hydration for Health. Published 2018.

8. Boschmann M, Steiniger J, Hille U, et al. Water-Induced Thermogenesis. *J Clin Endocrinol Metab.* 2003;88(12):6015–19.

9. Freeman S. How Water Works: Human Water Consumption. How Stuff Works. https://science.howstuffworks.com/environmental/earth/geophysics/h203.htm.

10. Pross N. Effects of Dehydration on Brain Functioning: A Life-Span Perspective. *Ann Nutr Metab.* 2017;70 Suppl 1:30–36.

11. Institute of Medicine. *Dietary Reference Intakes for Water, Potassium, Sodium, Chloride and Sulfate.* Washington, D.C.: The National Academies Press, 2005.

12. Guelinckx I, Tavoularis G, König J, Morin C, Gharbi H, Gandy J. Contribution of Water from Food and Fluids to Total Water Intake: Analysis of a French and UK Population Surveys. *Nutrients.* 2016;8(10).

13. Fedinick KP, Wu M, Panditharatne M, Olson ED. Threats on Tap: Widespread Violations Highlight Need for Investment in Water Infrastructure and Protections. Natural Resources Defense Council, 2017.

14. Environmental Working Group Tap Water Database. https://www.ewg.org/tapwater/.

Sharma S, Bhattacharya A. Drinking Water Contamination and Treatment Techniques. *Applied Water Science.* 2017;7:1043–67.

15. Fedinick KP, Wu M, Panditharatne M, Olson ED. Threats on Tap: Widespread Violations Highlight Need for Investment in Water Infrastructure and Protections. Natural Resources Defense Council, 2017.

16. Kilburn KH. Chlorine-Induced Damage Documented by Neurophysiological, Neuropsychological, and Pulmonary Testing. *Arch Environ Health.* 2000;55(1):31–37.

17. Postman A. The Truth About Tap: Lots of People Think Drinking Bottled Water Is Safer. Is It? Natural Resources Defense Council. https://www.nrdc.org/stories/truth-about-tap. Published 2016.

18. Postman A. The Truth About Tap: Lots of People Think Drinking Bottled Water Is Safer. Is It? Natural Resources Defense Council. https://www.nrdc.org/stories/truth-about-tap. Published 2016.

19. Leranth C, Hajszan T, Szigeti-Buck K, Bober J, MacLusky NJ. Bisphenol A Prevents the Synaptogenic Response to Estradiol in Hippocampus and Prefrontal Cortex of Ovariectomized Nonhuman Primates. *Proc Natl Acad Sci U.S.A.* 2008;105(37):14187–91.

20. Yang CZ, Yaniger SI, Jordan VC, Klein DJ, Bittner GD. Most Plastic Products Release Estrogenic Chemicals: A Potential Health Problem That Can Be Solved. *Environ Health Perspect.* 2011;119(7):989–96.

21. Brown KW, Gessesse B, Butler LJ, MacIntosh DL. Potential Effectiveness of Point-of-Use Filtration to Address Risks to Drinking Water in the United States. *Environ Health Insights.* 2017;11:1178630217746997.

22. United States Environmental Protection Agency. Safe Drinking Water Act: Consumer Confidence Reports (CCR). https://www.epa.gov/ccr. Published 2017.

23. Magro M, Corain L, Ferro S, et al. Alkaline Water and Longevity: A Murine Study. *Evid Based Complement Alternat Med.* 2016;2016:3084126.

24. Mantena SK, Jagadish, Badduri SR, Siripurapu KB, Unnikrishnan MK. In vitro Evaluation of Antioxidant Properties of Cocos nucifera Linn. Water. *Nahrung.* 2003;47(2):126–31.

25. Preetha PP, Devi VG, Rajamohan T. Hypoglycemic and Antioxidant Potential of Coconut Water in Experimental Diabetes. *Food Funct.* 2012;3(7):753–57.

26. Alleyne T, Roache S, Thomas C, Shirley A. The Control of Hypertension by Use of Coconut Water and Mauby: Two Tropical Food Drinks. *West Indian Med J.* 2005;54(1):3–8.

27. Sandhya VG, Rajamohan T. Comparative Evaluation of the Hypolipidemic Effects of Coconut Water and Lovastatin in Rats Fed Fat-Cholesterol Enriched Diet. *Food Chem Toxicol.* 2008;46(12):3586–92.

28. Feng L, Chong MS, Lim WS, et al. Tea Consumption Reduces the Incidence of Neurocognitive Disorders: Findings from the Singapore Longitudinal Aging Study. *J Nutr Health Aging.* 2016;20(10):1002–9.

29. Mancini E, Beglinger C, Drewe J, Zanchi D, Lang UE, Borgwardt S. Green Tea Effects on Cognition, Mood and Human Brain Function: A Systematic Review. *Phytomedicine.* 2017;34:26–37.

30. Gilbert N. The Science of Tea's Mood-Altering Magic. *Nature*. 2019;566(7742):S8–S9.

31. Kim J, Kim J. Green Tea, Coffee, and Caffeine Consumption Are Inversely Associated with Self-Report Lifetime Depression in the Korean Population. *Nutrients*. 2018;10(9).

32. Ohishi T, Goto S, Monira P, Isemura M, Nakamura Y. Anti-inflammatory Action of Green Tea. *Antiinflamm Antiallergy Agents Med Chem*. 2016;15(2):74–90.

33. Scholey A, Downey LA, Ciorciari J, et al. Acute Neurocognitive Effects of Epigallocatechin Gallate (EGCG). *Appetite*. 2012;58(2):767–70.

34. Chacko SM, Thambi PT, Kuttan R, Nishigaki I. Beneficial Effects of Green Tea: A Literature Review. *Chin Med*. 2010;5:13.

35. Gilbert N. The Science of Tea's Mood-Altering Magic. *Nature*. 2019;566(7742):S8–S9.

36. Oaklander M. Should You Drink Green Juice? *Time*. https://time.com/3818098/green-juice-kale-healthy/. Published 2015.

37. O'Callaghan F, Muurlink O, Reid N. Effects of Caffeine on Sleep Quality and Daytime Functioning. *Risk Manag Healthc Policy*. 2018;11:263–71.

38. Mojska H, Gielecińska I. Studies of Acrylamide Level in Coffee and Coffee Substitutes: Influence of Raw Material and Manufacturing Conditions. *Rocz Panstw Zakl Hig*. 2013;64(3):173–81.

Chapter 7: The Stress Offensive

1. Saad L. Eight in 10 Americans Afflicted by Stress. Gallup. https://news.gallup.com/poll/224336/eight-americans-afflicted-stress.aspx. Published 2017.

2. The American Institute of Stress. 42 Worrying Workplace Stress Statistics. https://www.stress.org/42-worrying-workplace-stress-statistics. Published 2019.

3. Xie L, Kang H, Xu Q, et al. Sleep Drives Metabolite Clearance from the Adult Brain. *Science*. 2013;342(6156):373–77.

4. Studte S, Bridger E, Mecklinger A. Nap Sleep Preserves Associative but Not Item Memory Performance. *Neurobiol Learn Mem*. 2015;120:84–93.

5. Okano K, Kaczmarzyk JR, Dave N, Gabrieli JDE, Grossman JC. Sleep Quality, Duration, and Consistency Are Associated with Better Academic Performance in College Students. *NPJ Sci Learn*. 2019;4:16.

6. National Sleep Foundation. How Lack of Sleep Impacts Cognitive Performance and Focus. https://www.sleepfoundation.org/articles/how-lack-sleep-impacts-cognitive-performance-and-focus.

7. Ben Simon E, Rossi A, Harvey AG, Walker MP. Overanxious and Underslept. *Nat Hum Behav*. 2020;4(1):100–110.

8. National Sleep Foundation. The Complex Relationship Between Sleep, Depression & Anxiety. https://www.sleepfoundation.org/excessive-sleepiness/health-impact/complex-relationship-between-sleep-depression-anxiety.

9. American Psychological Association. More Sleep Would Make Us Happier, Healthier and Safer. https://www.apa.org/action/resources/research-in-action/sleep-deprivation Published 2014.

10. Shi G, Xing L, Wu D, et al. A Rare Mutation of beta 1-Adrenergic Receptor Affects Sleep/Wake Behaviors. *Neuron*. 2019;103(6):1044–55 e1047.

11. Sheehan CM, Frochen SE, Walsemann KM, Ailshire JA. Are U.S. Adults Reporting Less Sleep?: Findings from Sleep Duration Trends in the National Health Interview Survey, 2004–2017. *Sleep*. 2019;42(2).

12. Lauderdale DS, Knutson KL, Yan LL, Liu K, Rathouz PJ. Self-Reported and Measured Sleep Duration: How Similar Are They? *Epidemiology*. 2008;19(6):838–45.

13. Peri C. 10 Things to Hate About Sleep Loss. WebMD. https://www.webmd.com/sleep-disorders/features/10-results-sleep-loss#1.

14. National Sleep Foundation. The Ideal Temperature for Sleep. https://www.sleep.org/articles/temperature-for-sleep/. Published 2020.

15. Koulivand PH, Khaleghi Ghadiri M, Gorji A. Lavender and the Nervous System. *Evid Based Complement Alternat Med*. 2013;2013:681304.

16. Hunter MR, Gillespie BW, Chen SY. Urban Nature Experiences Reduce Stress in the Context of Daily Life Based on Salivary Biomarkers. *Front Psychol*. 2019;10:722.

17. Hölzel BK, Carmody J, Vangel M, et al. Mindfulness Practice Leads to Increases in Regional Brain Gray Matter Density. *Psychiatry Res*. 2011;191(1):36–43.

18. Hölzel BK, Carmody J, Vangel M, et al. Mindfulness Practice Leads to Increases in Regional Brain Gray Matter Density. *Psychiatry Res*. 2011;191(1):36–43.

19. Brewer JA, Worhunsky PD, Gray JR, Tang YY, Weber J, Kober H. Meditation Experience Is Associated with Differences in Default Mode Network Activity and Connectivity. *Proc Natl Acad Sci U.S.A*. 2011;108(50): 20254–59.

20. Miller JJ, Fletcher K, Kabat-Zinn J. Three-Year Follow-Up and Clinical Implications of a Mindfulness Meditation-Based Stress Reduction Intervention in the Treatment of Anxiety Disorders. *Gen Hosp Psychiatry*. 1995;17(3):192–200.

21. Froeliger B, Garland EL, McClernon FJ. Yoga Meditation Practitioners Exhibit Greater Gray Matter Volume and Fewer Reported Cognitive Failures: Results of a Preliminary Voxel-Based Morphometric Analysis. *Evid Based Complement Alternat Med*. 2012;2012:821307.

22. Gotink RA, Vernooij MW, Ikram MA, et al. Meditation and Yoga Practice Are Associated with Smaller Right Amygdala Volume: The Rotterdam Study. *Brain Imaging Behav.* 2018;12(6):1631–39.

23. Streeter CC, Jensen JE, Perlmutter RM, et al. Yoga Asana Sessions Increase Brain GABA Levels: A Pilot Study. *J Altern Complement Med.* 2007;13(4):419–26.

24. Krishnakumar D, Hamblin MR, Lakshmanan S. Meditation and Yoga Can Modulate Brain Mechanisms That Affect Behavior and Anxiety—A Modern Scientific Perspective. *Anc Sci.* 2015;2(1):13–19.

25. Gothe NP, Hayes JM, Temali C, Damoiseaux JS. Differences in Brain Structure and Function Among Yoga Practitioners and Controls. *Front Integr Neurosci.* 2018;12:26.

26. Ma X, Yue ZQ, Gong ZQ, et al. The Effect of Diaphragmatic Breathing on Attention, Negative Affect and Stress in Healthy Adults. *Front Psychol.* 2017;8:874.

27. Steffen PR, Austin T, DeBarros A, Brown T. The Impact of Resonance Frequency Breathing on Measures of Heart Rate Variability, Blood Pressure, and Mood. *Front Public Health.* 2017;5:222.

28. Ma X, Yue ZQ, Gong ZQ, et al. The Effect of Diaphragmatic Breathing on Attention, Negative Affect and Stress in Healthy Adults. *Front Psychol.* 2017;8:874.

29. Lindgren L, Rundgren S, Winsö O, et al. Physiological Responses to Touch Massage in Healthy Volunteers. *Auton Neurosci.* 2010;158(1–2):105–10.

Chapter 8: Thinking Your Way to a Better Brain

1. Lee LO, James P, Zevon ES, et al. Optimism Is Associated with Exceptional Longevity in 2 Epidemiologic Cohorts of Men and Women. *Proc Natl Acad Sci U.S.A.* 2019;116(37):18357–62.

2. Comaford C. Got Inner Peace? 5 Ways to Get it Now. *Forbes.* https://www.forbes.com/sites/christinecomaford/2012/04/04/got-inner-peace-5-ways-to-get-it-now/#8232ec667275. Published 2012.

3. Millett M. Challenge Your Negative Thoughts. Michigan State University. https://www.canr.msu.edu/news/challenge_your_negative_thoughts. Published 2017.

4. Segerstrom S. The Structure and Consequences of Repetitive Thought. American Psychological Association. https://www.apa.org/science/about/psa/2011/03/repetitive-thought. Published 2011.

5. Watkins ER. Constructive and Unconstructive Repetitive Thought. *Psychol Bull.* 2008;134(2):163–206.

6. Sin NL, Graham-Engeland JE, Almeida DM. Daily Positive Events and Inflammation: Findings from the National Study of Daily Experiences. *Brain Behav Immun.* 2015;43:130–38.

7. Watkins ER. Constructive and Unconstructive Repetitive Thought. *Psychol Bull.* 2008;134(2):163–206.

8. Reynolds S. Happy Brain, Happy Life. *Psychology Today.* https://www.psychologytoday.com/us/blog/prime-your-gray-cells/201108/happy-brain-happy-life. Published 2011.

Mariën P, Manto M, eds. *The Linguistic Cerebellum.* New York: Academic Press, 2015.

9. Sapolsky RM. Stress and Plasticity in the Limbic System. *Neurochem Res.* 2003;28(11):1735–42.

10. Marchant NL, Howard RJ. Cognitive Debt and Alzheimer's Disease. *J Alzheimers Dis.* 2015;44(3):755–70.

11. Blackburn E, Epel E. *The Telomere Effect: A Revolutionary Approach to Living Younger, Healthier, Longer.* New York: Grand Central Publishing, 2017.

12. Neuvonen E, Rusanen M, Solomon A, et al. Late-Life Cynical Distrust, Risk of Incident Dementia, and Mortality in a Population-Based Cohort. *Neurology.* 2014;82(24):2205–12.

13. Goodin BR, Bulls HW. Optimism and the Experience of Pain: Benefits of Seeing the Glass as Half Full. *Current Pain and Headache Reports.* 2013;17(5):329.

14. Segerstrom SC, Taylor SE, Kemeny ME, Fahey JL. Optimism Is Associated with Mood, Coping, and Immune Change in Response to Stress. *J Pers Soc Psychol.* 1998;74(6):1646–55.

15. Chen L, Bae SR, Battista C, et al. Positive Attitude Toward Math Supports Early Academic Success: Behavioral Evidence and Neurocognitive Mechanisms. *Psychol Sci.* 2018;29(3):390–402.

16. Yanek LR, Kral BG, Moy TF, et al. Effect of Positive Well-Being on Incidence of Symptomatic Coronary Artery Disease. *Am J Cardiol.* 2013;112(8):1120–25.

17. Raghunathan R. How Negative Is Your "Mental Chatter"? *Psychology Today.* https://www.psychologytoday.com/us/blog/sapient-nature/201310/how-negative-is-your-mental-chatter. Published 2013.

18. Dispenza J. *You Are the Placebo: Making Your Mind Matter.* Carlsbad, CA: Hay House 2015.

19. Benedetti F, Carlino E, Pollo A. How Placebos Change the Patient's Brain. *Neuropsychopharmacology.* 2011;36(1):339–54.

20. Kirsch I, Deacon BJ, Huedo-Medina TB, Scoboria A, Moore TJ, Johnson BT. Initial Severity and Antidepressant Benefits: A Meta-analysis of Data Submitted to the Food and Drug Administration. *PLoS Med.* 2008;5(2):e45.

21. Vachon-Presseau E, Berger SE, Abdullah TB, et al. Brain and Psychological Determinants of Placebo Pill Response in Chronic Pain Patients. *Nat Commun.* 2018;9(1):3397.

22. Harvard Health Publishing. The Power of the Placebo Effect. https://www.health.harvard.edu/mental-health/the-power-of-the-placebo-effect. Published 2017.

23. Geers AL, Wellman JA, Fowler SL, Helfer SG, France CR. Dispositional Optimism Predicts Placebo Analgesia. *J Pain*. 2010;11(11):1165–71.

24. Kross E, Verduyn P, Demiralp E, et al. Facebook Use Predicts Declines in Subjective Well-Being in Young Adults. *PLoS One*. 2013;8(8):e69841.

25. Primack BA, Shensa A, Sidani JE, et al. Social Media Use and Perceived Social Isolation Among Young Adults in the U.S. *Am J Prev Med*. 2017;53(1):1–8.

26. Johnston WM, Davey GC. The Psychological Impact of Negative TV News Bulletins: The Catastrophizing of Personal Worries. *Br J Psychol*. 1997;88(Pt 1):85–91.

27. Sadeghi K, Ahmadi SM, Moghadam AP, Parvizifard A. The Study of Cognitive Change Process on Depression During Aerobic Exercises. *J Clin Diagn Res*. 2017;11(4):IC01–IC05.

28. Oppezzo M, Schwartz DL. Give Your Ideas Some Legs: The Positive Effect of Walking on Creative Thinking. *Journal of Experimental Psychology: Learning, Memory, and Cognition*. 2014;40(4):1142–52.

29. American Board of Professional Psychology. Search for Specialist. https://www.abpp.org/Directory.

Chapter 9: The Brain Games You Really Need

1. Wolinsky FD, Vander Weg MW, Howren MB, Jones MP, Dotson MM. A Randomized Controlled Trial of Cognitive Training Using a Visual Speed of Processing Intervention in Middle Aged and Older Adults. *PLoS One*. 2013;8(5):e61624.

2. Tennstedt SL, Unverzagt FW. The ACTIVE Study: Study Overview and Major Findings. *J Aging Health*. 2013;25(8 Suppl):3S–20S.

3. Jaeggi SM, Buschkuehl M, Jonides J, Perrig WJ. Improving Fluid Intelligence with Training on Working Memory. *Proc Natl Acad Sci U.S.A.* 2008;105(19):6829–33.

4. Nguyen T. 10 Proven Ways to Grow Your Brain: Neurogenesis and Neuroplasticity. Huffington Post. https://www.huffpost.com/entry/10-proven-ways-to-grow-yo_b_10374730. Published 2016.

5. Ballesteros S, Voelcker-Rehage C, Bherer L. Editorial: Cognitive and Brain Plasticity Induced by Physical Exercise, Cognitive Training, Video Games, and Combined Interventions. *Front Hum Neurosci*. 2018;12:169.

6. WebMD. Brain Exercises and Dementia. https://www.webmd.com/alzheimers/guide/preventing-dementia-brain-exercises#1. Published 2018.

7. Kidd DC, Castano E. Reading Literary Fiction Improves Theory of Mind. *Science*. 2013;342(6156):377–80.

Hurley D. Can Reading Make You Smarter? *Guardian*. https://www.theguardian.com/books/2014/jan/23/can-reading-make-you-smarter. Published 2014.

8. Berns GS, Blaine K, Prietula MJ, Pye BE. Short- and Long-Term Effects of a Novel on Connectivity in the Brain. *Brain Connect*. 2013;3(6):590–600.

9. Burmester A. Working Memory: How You Keep Things "In Mind" Over the Short Term. *Scientific American*. 2017.

10. Gernsbacher MA, Kaschak MP. Neuroimaging Studies of Language Production and Comprehension. *Annu Rev Psychol*. 2003;54:91–114.

11. Shah TM, Weinborn M, Verdile G, Sohrabi HR, Martins RN. Enhancing Cognitive Functioning in Healthly Older Adults: A Systematic Review of the Clinical Significance of Commercially Available Computerized Cognitive Training in Preventing Cognitive Decline. *Neuropsychol Rev*. 2017;27(1):62–80.

12. Roberts R, Kreuz R. Can Learning a Foreign Language Prevent Dementia? The MIT Press Reader. https://thereader.mitpress.mit.edu/can-learning-a-foreign-language-prevent-dementia/. Published 2019.

13. Alladi S, Bak TH, Duggirala V, et al. Bilingualism Delays Age at Onset of Dementia, Independent of Education and Immigration Status. *Neurology*. 2013;81(22):1938–44.

Bialystok E, Craik FI, Freedman M. Bilingualism as a Protection Against the Onset of Symptoms of Dementia. *Neuropsychologia*. 2007;45(2):459–64.

14. Bolwerk A, Mack-Andrick J, Lang FR, Dorfler A, Maihöfner C. How Art Changes Your Brain: Differential Effects of Visual Art Production and Cognitive Art Evaluation on Functional Brain Connectivity. *PLoS One*. 2014;9(7):e101035.

15. Chamberlain R, McManus IC, Brunswick N, Rankin Q, Riley H, Kanai R. Drawing on the Right Side of the Brain: A Voxel-based Morphometry Analysis of Observational Drawing. *Neuroimage*. 2014;96:167–73.

16. Carlson MC, Kuo JH, Chuang YF, et al. Impact of the Baltimore Experience Corps Trial on Cortical and Hippocampal Volumes. *Alzheimers Dement*. 2015;11(11):1340–48.

17. Carlson MC, Kuo JH, Chuang YF, et al. Impact of the Baltimore Experience Corps Trial on Cortical and Hippocampal Volumes. *Alzheimers Dement*. 2015;11(11):1340–48.

18. Piliavin JA, Siegl E. Health Benefits of Volunteering in the Wisconsin Longitudinal Study. *J Health Soc Behav*. 2007;48(4):450–64.

19. Sumowski JF, Rocca MA, Leavitt VM, et al. Reading, Writing, and Reserve: Literacy Activities Are Linked to Hippocampal Volume and Memory in Multiple Sclerosis. *Mult Scler*. 2016;22(12):1621–1625.

20. James KH, Engelhardt L. The Effects of Handwriting Experience on Functional Brain Development in Pre-literate Children. *Trends Neurosci Educ*. 2012;1(1):32–42.

21. Brooker H, Wesnes KA, Ballard C, et al. The Relationship Between the Frequency of Number-Puzzle Use and Baseline Cognitive Function in a Large Online Sample of Adults Aged 50 and Over. *Int J Geriatr Psychiatry*. 2019;34(7):932–40.

22. Maguire EA, Gadian DG, Johnsrude IS, et al. Navigation-Related Structural Change in the Hippocampi of Taxi Drivers. *Proc Natl Acad Sci U.S.A.* 2000;97(8):4398–403.

23. Parsons B, Magill T, Boucher A, et al. Enhancing Cognitive Function Using Perceptual-Cognitive Training. *Clin EEG Neurosci*. 2016;47(1):37–47.

Chapter 10: Biohacking Your Brain in Real Time

1. Centers for Disease Control and Prevention. Diabetes and Prediabetes. https://www.cdc.gov/chronicdisease/resources/publications/factsheets/diabetes-prediabetes.htm.

2. Centers for Disease Control and Prevention. Obesity and Overweight. https://www.cdc.gov/nchs/fastats/obesity-overweight.htm.

3. Wingo TS, Cutler DJ, Wingo AP, et al. Association of Early-Onset Alzheimer Disease with Elevated Low-Density Lipoprotein Cholesterol Levels and Rare Genetic Coding Variants of APOB. *JAMA Neurol*. 2019;76(7):809–17.

4. Parthasarathy V, Frazier DT, Bettcher BM, et al. Triglycerides are Negatively Correlated with Cognitive Function in Nondemented Aging Adults. *Neuropsychology*. 2017;31(6):682–88.

5. Reed B, Villeneuve S, Mack W, DeCarli C, Chui HC, Jagust W. Associations Between Serum Cholesterol Levels and Cerebral Amyloidosis. *JAMA Neurol*. 2014;71(2):195–200.

6. WebMD. What Is a C-Reactive Protein Test? https://www.webmd.com/a-to-z-guides/c-reactive-protein-test#1.

7. Brody JE. For Better Brain Health, Preserve Your Hearing. *New York Times*. https://www.nytimes.com/2019/12/30/well/live/brain-health-hearing-dementia-alzheimers.html. Published 2019.

8. Golub JS, Brickman AM, Ciarleglio AJ, Schupf N, Luchsinger JA. Association of Subclinical Hearing Loss with Cognitive Performance. *JAMA Otolaryngol Head Neck Surg*. 2019.

9. Deal JA, Reed NS, Kravetz AD, et al. Incident Hearing Loss and Comorbidity: A Longitudinal Administrative Claims Study. *JAMA Otolaryngol Head Neck Surg*. 2019;145(1):36–43.

10. Wolpert S. Dieting Does Not Work, UCLA Researchers Report. UCLA Newsroom. https://newsroom.ucla.edu/releases/Dieting-Does-Not-Work-UCLA-Researchers-7832. Published 2007.

Mann T, Tomiyama AJ, Ward A. Promoting Public Health in the Context of the "Obesity Epidemic": False Starts and Promising New Directions. *Perspect Psychol Sci*. 2015;10(6):706–10.

11. Norcross JC, Vangarelli DJ. The Resolution Solution: Longitudinal Examination of New Year's Change Attempts. *J Subst Abuse.* 1988;1(2):127–34.

12. Hills AP, Byrne NM, Lindstrom R, Hill JO. "Small Changes" to Diet and Physical Activity Behaviors for Weight Management. *Obes Facts.* 2013;6(3):228–38.

13. Harkin B, Webb TL, Chang BP, et al. Does Monitoring Goal Progress Promote Goal Attainment? A Meta-analysis of the Experimental Evidence. *Psychol Bull.* 2016;142(2):198–229.

14. Gordon ML, Althoff T, Leskovec J. Goal-Setting and Achievement in Activity Tracking Apps: A Case Study of MyFitnessPal. ACM International Conference on World Wide Web. https://cs.stanford.edu/people/jure/pubs/goals-www19.pdf. Published 2019.

15. Papalia Z, Wilson O, Bopp M, Duffey M. Technology-Based Physical Activity Self-Monitoring Among College Students. *Int J Exerc Sci.* 2018;11(7):1096–104.

16. Casey J. Body Fat Measurement: Percentage vs Body Mass. WebMD. https://www.webmd.com/diet/features/body-fat-measurement#1.

17. Kaviani S, vanDellen M, Cooper JA. Daily Self-Weighing to Prevent Holiday-Associated Weight Gain in Adults. *Obesity (Silver Spring).* 2019;27(6):908–16.

18. Sullivan AN, Lachman ME. Behavior Change with Fitness Technology in Sedentary Adults: A Review of the Evidence for Increasing Physical Activity. *Front Public Health.* 2016;4:289.

19. Hamblin J. The Futility of the Workout-Sit Cycle. *The Atlantic.* https://www.theatlantic.com/health/archive/2016/08/the-new-exercise-mantra/495908/. Published 2016.

20. Phillips P. *ASTD Handbook for Measuring and Evaluating Training.* Alexandria, Va.: ASTD Press; 2010.

21. Centers for Disease Control and Prevention. 5 Surprising Facts About High Blood Pressure. https://www.cdc.gov/features/highbloodpressure/index.html. Published 2016.

Epilogue: Biohacking Your Brain in the Twenty-First Century

1. Marins T, Rodrigues EC, Bortolini T, Melo B, Moll J, Tovar-Moll F. Structural and Functional Connectivity Changes in Response to Short-Term Neurofeedback Training With Motor Imagery. *Neuroimage.* 2019;194:283–90.

2. Harch PG, Fogarty EF. Hyperbaric Oxygen Therapy for Alzheimer's Dementia with Positron Emission Tomography Imaging: A Case Report. *Med Gas Res.* 2019;8(4):181–84.

3. Kjellgren A, Westman J. Beneficial Effects of Treatment with Sensory Isolation in Flotation-Tank as a Preventive Health-Care Intervention—A Randomized Controlled Pilot Trial. *BMC Complement Altern Med.* 2014;14:417.

4. Turner J, Gerard W, Hyland J, Nieland P, Fine T. Effects of Wet and Dry Flotation REST on Blood Pressure and Plasma Cortisol. In: Barabasz AF, et al., eds. *Clinical and Experimental Restricted Environmental Stimulation*. New York: Springer-Verlag, 1993.

5. Kjellgren A, Buhrkall H, Norlander T. Preventing Sick-Leave for Sufferers of High Stress-Load and Burnout Syndrome: A Pilot Study Combining Psychotherapy and the Flotation Tank. *International Journal of Psychology and Psychological Therapy*. 2011;11(2):297–306.

6. Jonsson K, Kjellgren A. Promising Effects of Treatment with Flotation-REST (Restricted Environmental Stimulation Technique) as an Intervention for Generalized Anxiety Disorder (GAD): a Randomized Controlled Pilot Trial. *BMC Complement Altern Med*. 2016;16:108.

7. Borrie RA. The Use of Restricted Environmental Stimulation Therapy in Treating Addictive Behaviors. *Int J Addict*. 1990–1991;25(7A–8A):995–1015.

8. Åsenlöf K, Olsson S, Bood SA, Norlander T. Case Studies on Fibromyalgia and Burn-Out Depression Using Psychotherapy in Combination with Flotation-Rest: Personality Development and Increased Well-Being. *Imagination, Cognition and Personality*. 2007;26(3):259–71.

9. Jiang H, White MP, Greicius MD, Waelde LC, Spiegel D. Brain Activity and Functional Connectivity Associated with Hypnosis. *Cereb Cortex*. 2017;27(8):4083–93.

INDEX

ABOUT THE AUTHORS

Kristen Willeumier, Ph.D., is a neuroscientist with laboratory training in neuroendocrinology, neurophysiology, and neurogenetics. She earned a B.A. in psychology from Boston College, an M.S. in physiological science from UCLA, and an M.S. and Ph.D. in neurobiology from the UCLA David Geffen School of Medicine. She also completed doctoral training in the Department of Neurology at Cedars-Sinai Medical Center in Los Angeles. Dr. Willeumier was formerly the director of neuroimaging research for the Amen Clinics, a nationally recognized outpatient health-care clinic, during which time she conducted groundbreaking research on understanding and treating brain trauma in professional NFL players. She has been published in peer-reviewed journals, including *The Journal of Neuroscience*, the *Journal of Alzheimer's Disease*, and *Translational Psychiatry*. Dr. Willeumier appears frequently on various media outlets to speak about brain health and neuroscience.

Sarah Toland is a longtime journalist and coauthor of several books, including the *New York Times* bestselling *Strong Is the New Beautiful* (2014) with Olympic skier Lindsey Vonn and the national bestselling *The Self-Care Solution* (2019) with Jennifer Ashton. She has written for the *New York Times, Sports Illustrated, Men's Journal,* and *Prevention,* among other publications, and she has appeared often as an independent health expert on national television.